W9-AQV-217

EINSTEIN'S
MONSTERS

ALSO BY CHRIS IMPEY

Humble Before the Void

Shadow World

Dreams of Other Worlds

How It Began

How It Ends

The Living Cosmos

Beyond

EINSTEIN'S MONSTERS

The Life and Times of Black Holes

CHRIS IMPEY

W. W. NORTON & COMPANY / *Independent Publishers Since 1923* / New York · London

For information about permission to reproduce selections from this book, write to
Permissions, W. W. Norton & Company, Inc., 500 Fifth Avenue, New York, NY 10110

For information about special discounts for bulk purchases, please contact
W. W. Norton Special Sales at specialsales@wwnorton.com or 800-233-4830

Manufacturing by LSC Communications Harrisonburg
Book design by Ellen Cipriano
Production manager: Lauren Abbate

Library of Congress Cataloging-in-Publication Data

Names: Impey, Chris, author.
Title: Einstein's monsters : the life and times of black holes / Chris Impey.
Other titles: Life and times of black holes
Description: First edition. | New York : W.W. Norton & Company, [2019] |
Includes bibliographical references and index.
Identifiers: LCCN 2018019192 | ISBN 9781324000938 (hardcover)
Subjects: LCSH: Black holes (Astronomy—Popular works. |
Gravitation—Popular works.
Classification: LCC QB843.B55 I47 2019 | DDC 523.8/875—dc23
LC record available at https://lccn.loc.gov/2018019192

W. W. Norton & Company, Inc., 500 Fifth Avenue, New York, N.Y. 10110
www.wwnorton.com

W. W. Norton & Company Ltd., 15 Carlisle Street, London W1D 3BS

1 2 3 4 5 6 7 8 9 0

To Dinah,
My love and inspiration.

*Stars, hide your fires; Let not light see
my dark and deep desires.*

WILLIAM SHAKESPEARE,
MACBETH, ACT 1, SCENE 4

CONTENTS

ACKNOWLEDGMENTS

I'm grateful to my wife, Dinah, for support of all my creative endeavors. Thanks go to my agent, Anna Ghosh, for steering my writing in fruitful directions. It's been a pleasure to work with Tom Mayer, my editor at Norton, and I appreciate Sarah Bolling's comments on the first draft. I acknowledge the Aspen Center for Physics for its atmosphere of tranquility and intellectual stimulation that is so conducive to science writing. I have benefited from many conversations about black holes with my colleagues at the University of Arizona and around the world. Their excitement reminds me that the universe is a wondrous place. It's a privilege to be a scientist and educator and share that excitement with others.

FOREWORD

Black holes are the best known and least understood objects in the universe. The term is used colloquially to talk about an entity that sucks in everything around it. Black holes appear onscreen and in fiction; they've been co-opted by pop culture. A black hole is shorthand for something enigmatic, with a sinister edge. I call them "Einstein's monsters" as a metaphor. They are powerful and beyond anyone's control. Einstein did not create black holes, but he developed the best theory of gravity that we have for understanding them.[1]

What most people think they know about black holes is wrong. They're not cosmic vacuum cleaners, sucking in everything in the vicinity. They only distort space and time very close to their event horizons. Black holes represent a small fraction of the mass of the universe, and the nearest examples are several hundred trillion miles away. It's unlikely that they can be used to time-travel or visit other universes. Black holes aren't even black. They emit a fizz of particles and radiation and most are part of binary systems where gas falling in heats up and glows intensely. Black holes aren't necessarily hazardous. You could fall into the black hole at the center of most galaxies and not feel a thing, although you'd never get to tell anyone what you saw.

This book is an introduction to black holes, large and small. Black holes are deceptively simple, but the math needed to understand them is fiendishly complex. We'll meet scientists who revealed black holes to humankind, from theorists who dared dream of dark stars hundreds of years ago to those who wrestled with general relativity, and beyond.

Black holes cannot be understood without Einstein's theory of general relativity, developed a century ago, which says that space and time are distorted by matter. In the extreme case where mass is highly concentrated, a region of space is "pinched off" from the rest of the universe and nothing can escape, not even light. That's a black hole. But even Einstein was skeptical of their reality. He was not alone; many notable physicists doubted that they existed.

They do exist. Evidence has accumulated for forty years that when massive stars die, no force in nature can resist the gravitational collapse of their cores. A gas ball 10 times the size of the Sun crunches down to a dark object the size of a small town. More recently, it's become clear that the center of every galaxy contains a massive black hole, ranging in mass over a factor of a billion.

Examining where black holes live, we'll learn about binary systems, where a black hole is in a gravitational waltz with a normal star. We'll see that the best evidence for any black hole is at the center of our own galaxy, where dozens of stars swarm like angry bees around a dark object 4 million times the mass of the Sun. When the massive black holes that all galaxies harbor rouse from their slumber and begin feeding, they can be seen across distances of billions of light years. These gravitational engines are the most powerful sources of radiation in the universe.

Recently, physicists have learned to see with "gravity eyes" by detecting gravitational waves. When two black holes collide, they release a cymbal clash of ripples in space-time that race outward at the speed of light and contain information about the violent encounter. A new window has opened onto black holes and all situations where gravity is strong and changing. Gravitational waves provide unequivocal proof, if any were still needed, that nature makes black holes. Every five minutes, a pair of black holes merges somewhere in the universe, pouring gravitational waves out into space.

We're far from knowing everything there is to know about black holes. They continue to surprise and delight. Black holes allow general relativity to be tested in new ways. Nobody knows if these tests will affirm the theory or lead to its downfall. There is vigorous debate over information loss in black holes and whether or not the information is somehow coded on the event horizon. Theorists hope that black holes may be places where string theory can be verified, finally realizing Einstein's quest to unify quantum mechanics and general relativity.

This book has two parts. The first covers the evidence we have that black holes exist. This evidence spans a range of black holes, from those not much more massive than the Sun to leviathans the mass of a small galaxy. The second explains how black holes are born and die. It also explains how black holes push our theories of nature to the limit. Along with the stories of black holes, there are personal stories, including some of my own, as a reminder that while science is dispassionate, scientists are made of flesh and blood, with all the flaws and foibles that entails. As I am covering research in a fast-moving subject, some results quoted here may not stand the test of time. Any resulting errors, omissions, or misrepresentations are mine alone.

We can imagine that intelligent creatures on many of the trillions of habitable worlds in the cosmos have deduced the existence of black holes. Perhaps some have learned how to make them and harness their power. Humans are a young species, but we can be proud to be members of the special club that knows of black holes.

Chris Impey
Tucson, Arizona
April 2018

PART A

Evidence for Black Holes, Large and Small

How did scientists arrive at the concept of a black hole? In this part of the book, we'll see how speculation started after Newton proposed his theory of gravity, and spread after Einstein articulated the implications of his theory of general relativity. Today we know black holes have two key ingredients: an event horizon, which acts as an information barrier, and a singularity, or central point of infinite mass density. Many notable physicists, including Einstein himself, rebelled against such a bizarre state of matter. But others showed that a massive star core would collapse to a density from which particles and radiation could not escape.

If theorists had trusted the beauty of the mathematics of general relativity, they would have had no reason to doubt the existence of black holes. But science is empirical, so astronomers labored to track down their elusive quarry. It was only with the advent of X-ray astronomy, a decade after Einstein's death, that researchers could see the hot accretion disk and twin jets that form when a black hole absorbs gas from the universe surrounding it. Hunting dark and dead stars is challenging. Even after fifty years of work, there are only three

dozen stellar corpses proven to be black holes beyond a reasonable doubt. They are the closest examples of a projected
population of roughly 10 million scattered in the Milky
Way. As this evidence was being accumulated painstakingly,
astronomers made the surprising discovery that massive black
holes lurk in the centers of galaxies. When these black holes
are consuming matter, they become the brightest objects in
the universe.

1.

THE HEART OF DARKNESS

SCIENTISTS ARE OPTIMISTS. They're impressed by the reach and predictive power of theories like relativity and natural selection. They believe that the rapid progress of physics, astronomy, and biology observed over the last few decades will continue, and that science will spread its explanatory tentacles ever farther into the natural world.

But what if this ambition were to hit an immovable obstacle? What if the cosmos were to harbor objects that resist our prying eyes, objects that are ciphers? Worse still, what if these mysterious entities were predicted by our best physical theories yet had properties that cast doubt on those theories? Welcome to the world of black holes.

An English Clergyman Imagines Dark Stars

John Michell was, according to his contemporaries, "a little short man, of a black complexion, and fat." He spent most of his adult life as the rector of a church in a small town in northern England. Yet famous thinkers of the day like Joseph Priestley, Henry Cavendish, and Benjamin Franklin beat a path to his door, because Michell was also a versatile

and accomplished scientist. His modesty, and his quiet life as a clergyman, have caused him to be neglected by history.

Michell studied mathematics at Cambridge University, where he
later taught mathematics, Greek, and Hebrew. He founded the field of
seismology by recognizing that earthquakes travel as waves through the
Earth, an insight which earned him a place in the Royal Society. It was
Michell who devised the experimental equipment that was later used
by Henry Cavendish to measure the gravitational constant, the fundamental number that underpins all gravity calculations. He also was the
first to apply statistical methods to astronomy, arguing that many pairs
and groups of stars in the night sky must be physically associated rather
than chance alignments.[1]

The clergyman was at his most visionary in proposing that some
stars might have gravity so strong that not even light could escape. He
introduced the idea in a 1784 paper with the unwieldy title "On the
Means of Discovering the Distance, Magnitude, etc. of the Fixed Stars,
in Consequence of the Diminution of the Velocity of Their Light, in
Case Such a Diminution Should be Found to Take Place in any of Them,
and Such Other Data Should be Procured from Observations, as Would
be Further Necessary for That Purpose."[2]

The gist of the paper can be explained in not many more words than
the title. Michell understood the idea of escape velocity and the fact that
it would be determined by the mass and size of a star. He followed Isaac
Newton in believing light to be a particle, and reasoned that light would
be slowed down by a star's gravity. He wondered what would happen
if a star was so massive and its gravity so strong that its escape velocity
was equal to the speed of light, and hypothesized many "dark stars" that
were undetectable because light couldn't escape from them.[3]

Michell's reasoning was flawed, but only because he was working
with Newtonian physics. In 1887, Albert Michelson and Edward Morley
demonstrated that light always travels at the same speed, regardless of
the Earth's motion.[4] Not until 1905 did Einstein make this result the

premise of his special theory of relativity, proposing that the speed of light doesn't depend on the local strength of gravity. Michell also erred in conceiving of dark stars as being 500 times larger than the Sun but having the same density. Stars that massive don't exist. The extreme effects of gravity are only realized when the density is high, which happens when a star like the Sun has been compressed into a tiny volume.

A Great French Mathematician Weighs In

A dozen years after Michell's speculation about dark stars, French scientist and mathematician Pierre-Simon Laplace wrote about the same topic in his book *Exposition of the System of the World*. More celebrated than Michell, Laplace was president of the Institut Français, acted as an advisor to Napoleon, and became a count and later a marquis. Like Michell, he studied theology and was from a religious family, but the call of mathematics was louder than the call of God.

Laplace was apparently unaware of Michell's work. In a two-volume treatise on astronomy, Laplace makes brief mention of the idea of a dark star as he considers the gravity of a hypothetical star much larger than the Sun: "It is therefore possible that the largest luminous bodies in the universe may, through this cause, be invisible." A colleague challenged Laplace to provide mathematical proof of this, which he did, three years later, in 1799.[5] His work was flawed for the same reason as Michell's. The densest substance known at the time was gold, 5 times denser than the Earth and 14 times denser than the Sun. It may have been hard for scientists to conceive of states of matter millions of times denser, as is required for our modern understanding of a black hole (Figure 1). Laplace excised any reference to dark stars from later editions of his book, probably because Thomas Young showed in 1799 that light behaves like a wave and it seemed unlikely that gravity could slow down a wave.

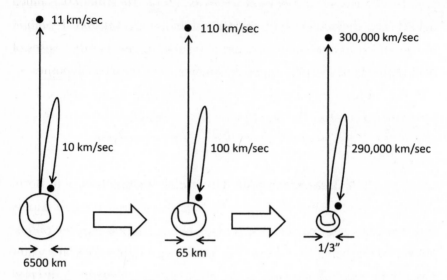

FIGURE 1. The concept of a black hole based on Newtonian gravity. The escape velocity of the Earth is 11 km/s; any object launched with that speed will escape Earth's gravity. If the Earth shrinks by a factor of 100, the escape velocity rises to 110 km/s. A black hole would result if the Earth could be compressed to a radius of 1/3 inch, where the escape velocity is the speed of light. *John D. Norton/University of Pittsburgh*

The concept of black holes couldn't emerge fully without a new theory of gravity. Newton's theory is simple: space is smooth and linear and stretches infinitely in all directions. Time is smooth and linear and flows into the infinite future. Space and time are distinct and independent. Stars and planets move through empty space governed by a force that depends on their masses and the distances between them. This is Newton's elegant universe.[6]

Richard Westfall, Newton's biographer and a brilliant scholar in his own right, said, "The end result of my study of Newton has served to convince me that with him there is no measure. He has become for me wholly other, one of the tiny handful of supreme geniuses who have shaped the categories of the human intellect, a man not finally reducible to the criteria by which we comprehend our fellow beings."[7] Yet even Newton's great intellect did not fully illuminate gravity. He couldn't explain how it operates instantaneously and invisibly across a vacuum. He admitted as much in his masterwork on gravity from 1687, *Philosophiae Naturalis Principia Mathematica*. He wrote, "I have not been able to discover the causes of those properties of gravity from phenomena, and I frame no hypotheses."

Understanding the Fabric of Space-Time

Albert Einstein, a twenty-six-year-old clerk at the patent office in Bern, dismantled the Newtonian system. In 1905, Einstein wrote four papers that would change the face of physics.[8] In one, he looked at the photoelectric effect, where electrons are released when light shines on a material. He argued that light acts like a particle, carrying energy in discrete amounts called quanta. It was this work, rather than his more famous theories of relativity, that won him a Nobel Prize (Figure 2). Experi-

FIGURE 2. Albert Einstein in 1921, five years after he published his general theory of relativity. His theory is a radical departure from Newton's gravity, which is based on linear and absolute space and time. In general relativity, space-time has curvature caused by the mass and energy it contains. *Ferdinand Schmutzer*

ments by Thomas Young and others had firmly established that light manifests behaviors of diffraction and interference, so physicists were forced to accept that light is somehow both wave-like and particle-like.

Another short paper presented the most famous equation in physics: $E = mc^2$. This says that mass and energy are equivalent and interchangeable. Since the speed of light, c, is a very large number, a tiny amount of mass can be converted into an enormous amount of energy. Mass is like a "frozen" form of energy, which is why nuclear weapons are so powerful. Conversely, energy has a tiny amount of equivalent mass. Given this equation, it makes sense that photons are affected by gravity.

A third paper laid out the special theory of relativity. The theory built on Galileo's idea that the laws of nature should be the same for all observers moving at constant speed relative to each other, and added a second premise: that the speed of light doesn't depend on motion of an observer. The second premise is radical, as a thought experiment will show.[9] You shine a flashlight at someone far away. They measure the photons arriving at 300,000 km/s, the speed of light. Suppose you rush toward them at half the speed of light. They still see the photons arrive at the same speed, and not at 450,000 km/s. Now suppose you rush away from them at half the speed of light. They again see the photons arrive at the same speed, not at 150,000 km/s. Light does not obey simple arithmetic. Light speed is a universal constant, and the implications of this are profound. Speed is distance divided by time; if speed is constant, then space and time must be supple. As objects travel very fast and approach the speed of light, they shrink in the direction of motion and their clocks run slower. Einstein's theory says that light is the fastest thing there is, so he also predicted that objects will get more massive as they approach the speed of light, increasing their inertia so they are never able to reach or exceed that speed.

As remarkable as all this work was, Einstein was just flexing his muscles for his seminal achievement, the general theory of relativity. With general relativity, Einstein extended his ideas from constant motion to accelerated motion, and brought gravity into the mix. He began with another Galilean insight. The Renaissance polymath had shown that all objects fall at the same rate regardless of mass. This means that inertial mass (the resistance of an object to a change in its motion) is the same as its gravitational mass (how an object responds to the force of gravity). This was a coincidence and a puzzle to Galileo, but Einstein suspected it was the key to a new concept of gravity.

Imagine you're in a closed elevator stuck on the ground floor. You feel your normal weight; anything you drop accelerates at 9.8 m/s². It's a familiar gravity situation. Now imagine you're in a sealed box in space (which looks like the inside of an elevator) being accelerated by a spaceship at 9.8 m/s². One situation involves gravity and the other doesn't. But Einstein realized there was no experiment that could distinguish them (Figure 3). Here are two more situations. In one, you're trapped inside

FIGURE 3. In general relativity, there's no difference between acceleration due to gravity and acceleration due to any other force. So someone in a rocket accelerating in deep space at 9.8 m/s/s (left) will feel as if they are experiencing Earth's gravity and not notice any difference in the behavior of falling objects relative to being stationary on the Earth's surface. *Markus Pössel*

the elevator in deep space. Weightless, you float inside the elevator. In the other, the elevator is in a tall building and the cable has broken so it's plunging to the bottom of the shaft. There is no way to distinguish these situations either. Gravity is indistinguishable from any other force. This "equivalence principle" is central to Einstein's general theory of relativity. Despite the calamity implied by a drastic plunge in an elevator, Einstein said it was his "happiest thought" that a person falling would not feel their weight.

Einstein's new conception of gravity was geometric. The equations of general relativity relate the amount of mass and energy in a region to the curvature of the space. The flat and linear space of Newton, with objects contained within it, is replaced by space that's curved by the objects it contains (Figure 4).[10] Space and time are linked, so gravity can distort time as well as space. Physicist John Wheeler, whom we'll meet later as the person who coined the term "black hole," put it succinctly: "Matter tells space how to curve. Space tells matter how to move." Let's also hear it in the words of the poet Robert Frost, who had an ambivalent relationship with the revelations of relativity. In the sonnet "Any Size We Please," he found the thought of infinite space terrifying but took comfort in the curvature that described a black hole:

> He thought if he could have his space all curved,
> Wrapped in around itself and self-befriended,
> His science needn't get him so unnerved.
> He had been too all out, too much extended.
> He slapped his breast to verify his purse
> And hugged himself for all his universe.[11]

Three of the effects of general relativity are particularly relevant for situations of dense matter, epitomized by black holes. First is the deflection of light as it follows the undulations of space-time due to concentrations of mass. This was the initial, classic test of general rela-

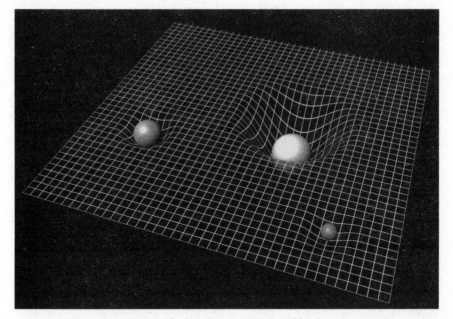

FIGURE 4. In Einstein's theory of relativity space is curved by the mass it contains, seen here in a two-dimensional analogy, where the curvature increases with mass. A black hole is a situation where space-time is "pinched off" from the rest of the universe. In a situation of normal stars and planets, the space distortion is almost imperceptible. *ESA/Christophe Carreau*

tivity, carried out in 1919, three years after Einstein published the theory. A team led by the great English astrophysicist Arthur Eddington measured the slight bending of starlight as it passed near the edge of the Sun. It wasn't a very precise measurement, but the affirmation of relativity turned Einstein into a celebrity and vaulted him into the science stratosphere. In 1995, a more accurate measurement agreed with Einstein's prediction to 0.01%.[12]

A second effect is a loss of energy as light leaves a massive object, called gravitational redshift. Think of it as photons struggling against gravity. The effect was first measured experimentally in 1960. A closely related effect is time dilation, the prediction that clocks run slower when gravity is stronger. Time dilation was first detected in 1971, when an atomic clock was flown at high altitude and it ran very slightly faster than an identical atomic clock kept on the ground. In 2010, the time dilation was detected over a vertical separation of just one meter, which required a clock with the amazing precision of one second over 4 billion years.[13] Time dilation measurements also agree with predictions from theory to an accuracy of 0.01%. General relativity has passed all its experimental tests with flying colors.

General relativity seems esoteric and remote from everyday life, but the GPS system would fail totally if time dilation was not included in the calculations. Locating a phone on the Earth to within a meter depends on extremely accurate measurements of orbiting satellites containing atomic clocks.[14] Relativity calculations are done by the computer chips in your cell phone; without these corrections GPS would be off by 10 kilometers after one day. The effects of relativity are subtle in the Solar System and where gravity is weak, but we'll see that they're turbocharged when stars collapse and gravity is strong.

A Singularity and a Life Cut Short

General relativity is an austere and beautiful theory. Einstein said, of his creation, "Scarcely anyone who fully understands this theory can escape from its magic."[15] But few people have the mathematical fortitude to come to grips with relativity. In its most compact form, a single equation relates the density of mass-energy to the curvature of space-time. That's like five-minute Shakespeare. The full performance is a set of ten coupled, nonlinear, hyperbolic-elliptic, partial differential equations. The underlying mathematics is based on manifolds, complex multidimensional shapes that are to Euclidean space as an origami dragon is to a flat sheet of paper.[16]

Einstein worked out approximate solutions to his theory so that Arthur Eddington could set up his expedition to measure the gravitational deflection of starlight during an eclipse. He doubted that the equations could be solved exactly, but general relativity immediately attracted attention from the best minds in physics. One of those people made remarkable progress. Karl Schwarzschild was born in Frankfurt and became a precocious student, publishing two papers on binary star orbits when he was sixteen. He rose quickly to become a professor and director of the observatory at the University of Göttingen. Though he was over forty when World War I broke out, patriotism inspired him to enlist in the German military. He served on the western and eastern fronts and rose to the rank of lieutenant in the artillery.

Schwarzschild corresponded with Einstein in late 1915 while enduring bitter cold at the Russian front. "The war treated me kind enough," Schwarzschild wrote, "in spite of the heavy gunfire, to allow me to get away from it all and take this walk in the land of your ideas."[17] Einstein was impressed by Schwarzschild's exact solution to his equations and presented it to the German Academy of Sciences. However, a rare and painful skin condition called pemphigus prevented Schwarzschild from pursuing his ideas. He submitted a paper for publication in February 1916, was sent home from the Russian front in March, and died in May.

What was the solution that Schwarzschild found? That escape veloc-

FIGURE 5. Black holes are simple objects, characterized by mass and spin. The event horizon is an information barrier that separates regions of space-time we can see and those we cannot. The singularity at the center of a black hole is a point of infinite mass density. *Monica Turner/Science in School/EIROforum*

ity from the surface of an object depends on its mass and its radius. Michell and Laplace had speculated about the possibility of light being trapped by a large, massive star with the same density as the Sun. Schwarzschild realized the escape velocity could also reach the speed of light if a star like the Sun collapsed to a high density. His solution had two striking features. The first was that gravity collapses the object to a state of infinite mass density called a singularity. The second was the prediction of a gravitational boundary that would forever trap what was inside. This is the event horizon. A singularity and an event horizon are the two essential ingredients of a black hole (Figure 5).

The Master of Implosions and Explosions

Einstein wasn't pleased. Both he and Eddington were convinced that a singularity was a sign of imperfect physical understanding. It made no sense for a physical object to have zero size and infinite mass density. Einstein's theory had created something monstrous. Other physicists thought Schwarzschild's solution was an esoteric curiosity. For a star like the Sun, the Schwarzschild radius—that is, the size of the event horizon—was 3 kilometers. How could a star that was 1.4 million kilometers across—over 100 times the size of the Earth— collapse to the size of a village?

Another wunderkind of physics was convinced it was possible. Robert Oppenheimer was born in New York and studied physics at Harvard. After getting his PhD, he traveled around Europe and steeped himself in the emerging field of quantum mechanics. His scientific interests were voracious. Among other accomplishments, he was the first to apply quantum theory to molecules, he predicted antimatter, and he pioneered a theory of cosmic rays. Along the way, he built the best theoretical physics program in the world at the University of California at Berkeley. Oppenheimer was a cultured man with a deep interest in art and music, who had studied Sanskrit and read philosophy in the original ancient Greek. He had a strong social conscience and left-wing leanings.[18]

Oppenheimer developed the tools for understanding nuclear matter and realized that astrophysics provided some exotic real-world examples. As a star evolves it keeps a delicate balance between gravity, which is always pulling inward, and pressure from fusion reactions, which pushes outward. The Sun is stable and it has a constant size as long as nuclear reactions continue. When the Sun runs out of hydrogen fuel it will collapse to a dense state of matter supported by a quantum mechanical force called degeneracy pressure. That's called a white dwarf. Indian astrophysicist Subrahmanyan Chandrasekhar worked out that the gravity of a star more massive than the Sun could overcome the force of degeneracy pressure and collapse down to the density of a vast atomic nucleus. That's called a neutron star. In 1939, Oppenheimer and one of his graduate students wrote a paper called "On Continued Gravitational Contraction,"[19] wherein, with a challenging calculation, they showed that an even more massive star will collapse until it reaches a density beyond any known form of matter. At the end of the massive star's life, a black hole inevitably forms.

In 1942, Oppenheimer was tapped to lead the American effort to develop an atomic bomb. He assembled a dream team of talented physicists to work at a secret site in Los Alamos in northern New Mexico, in an intense effort aimed at gaining decisive advantage in the war against Japan.[20] Oppenheimer was committed to the work but showed glimmers of conflict. After witnessing the Trinity test explosion in 1945, he simply said to his brother, "it worked." Later, he chose the famous words from the *Bhagavad Gita*, "I am become death, destroyer of worlds."[21] After the war, Oppenheimer's politics led to his downfall. He was subjected to the humiliation of an anti-Communist witch hunt and stripped of his security clearance. His reputation was never fully rehabilitated. But among other landmarks in his massive physics legacy, he had taken black holes from speculation to plausibility.

Coining the Perfect Term for the Inscrutable

Physicists don't always get along. The greatest scientists are often highly competitive, and they are also passionate about understanding how nature works. I've witnessed fierce rivalries in my field, and I've flinched at the brutal words scientists sometimes use against one another. Usually best ideas are affirmed and hard feelings are set aside. But sometimes conflict is rooted in personality, as was the case with Robert Oppenheimer and John Wheeler, the man who coined the term "black hole" (Figure 6).

Wheeler was mentored by the great Danish physicist Niels Bohr, who instilled in him a habit of not just cranking through the equations but also asking deep questions about the nature of reality as revealed by physics. He considered but decided against doing his PhD thesis work in Berkeley with Oppenheimer, a man only seven years his senior. Wheeler spent most of his career as a professor at Princeton, where he mentored many of the best physicists of the second half of the twentieth century. He deserves much of the credit for making the study of gravity a legitimate, mainstream subject. In 1973, as he neared retirement, Wheeler and two former students wrote the landmark text *Gravitation,* which physics graduate students still read today.[22]

On the same day in 1939 that Oppenheimer's paper on stellar collapse was published, Wheeler and Bohr published the explanation for nuclear fission—while over in Europe, Hitler invaded Poland. Like Einstein and Eddington before him, Wheeler rejected the idea of a singularity; for him, too, it was a violation of physics. At a conference in 1958, Wheeler gave a talk in which he rejected Oppenheimer's idea, saying, "It does not give an acceptable answer." A spirited debate followed. Oppenheimer was often intense, impatient, and personally aloof. Wheeler was earnest, engaged, and curious to learn about everyone he met. Wheeler said of Oppenheimer, "I never really understood him. I always felt that I had to have my guard up." (Wheeler did come around to Oppenheimer's idea after computer codes used to model bombs showed it was plausible, and at a conference in 1962, he praised Oppenheimer's work. Oppenheimer, however, did not hear Wheeler's words of support, as he had chosen to stay outside the hall talking to a colleague.[23])

Their animus was fueled by a major divergence of opinion during the war. Oppenheimer was the chief architect of the atomic bomb program that helped end the war, but after that he put his energy into nonproliferation. Meanwhile, Wheeler and Edward Teller led the effort to build the much more powerful hydrogen bomb, which they nicknamed the "Super."[24] Oppenheimer opposed them, saying, "Let Teller and Wheeler go ahead. Let them fall on their faces."[25] But they did not, and Oppenheimer later bowed to their technical prowess, which made a fusion bomb possible. For his part, Wheeler became hawkish after the death of his brother in 1944 in action in Italy. He bitterly regretted that the bomb hadn't been developed in time to change the course of the war in Europe.

FIGURE 6. John Archibald Wheeler, one of the foremost physicists of the second half of the twentieth century and author of the classic textbook *Gravitation*. He coined the term "black hole." *Office of Public Records, The Dolph Briscoe Center for American History, The University of Texas at Austin*

In a talk in 1967, Wheeler remarked that after you say "gravitation-ally completely collapsed objects" enough times, you start to look for a better name. Someone in the audience (who has never been identified) called out, "How about black hole?" Wheeler began to use it to see if it would catch on. It did. Like the term "big bang," which was also coined by someone who did not subscribe to the idea, "black hole" is colloquial yet accurate.[26] In his autobiography, Wheeler wrote that a black hole "teaches us that space can be crumpled like a piece of paper into an infinitesimal dot, that time can be extinguished like a blown-out flame, and that the laws of physics that we regard as sacred, as immutable, are anything but."

A Genius Struggles with Gravity and Disease

Stephen Hawking was another brilliant mind who took on the challenge of black holes. His story is so familiar we almost forget to be amazed. A diffident and mediocre student at school, he scraped a First Class Hon-ors degree after working no more than an hour a day for three years. Stricken by ALS, a degenerative motor neuron disease, at the age of twenty-one, he was given two years to live. Yet he was elected to the Royal Society at age thirty-two, and to the Lucasian Chair of Mathe-matics at Cambridge University—once held by Isaac Newton—at age thirty-five. He nearly died of pneumonia in the 1980s, which resulted in him losing his speech and being given his now iconic mechanical voice. A Brief History of Time launched him as a celebrity and sold over 10 million copies.[27] By the time of his death in March 2018, he had outlived his original death sentence by more than half a century (Figure 7).

Those close to Hawking described a personality with sharp edges,[28] but in physics, at least, he was the most brilliant and original mind since Einstein.[29] In his PhD thesis, Hawking focused on a topic that most physicists preferred to avoid: singularities. As we've seen, the implica-

FIGURE 7. Stephen Hawking is momentarily the master of gravity during a flight on a modified Boeing 727 aircraft in 2007. Space entrepreneur Peter Diamandis arranged the flight with the cooperation of NASA. Hawking had hoped to have a longer zero-gravity flight with Virgin Galactic. *Jim Campbell/Aero-News Network*

tion of a singularity at the center of a black hole caused even Einstein to doubt his own theory. In mathematics, a singularity is a situation in which a function has an infinite value. This happens all the time and it isn't fatal; mathematics has many ways to manipulate and deal with infinities. In physics, however, infinity is a big problem. For example, a theory describing liquids might predict that under some conditions the density of a liquid becomes infinite. That's clearly nonphysical and indicates a deficiency in the theory. Hawking wasn't so sure that singularities indicated a problem with general relativity. He formed a collaboration with the Oxford mathematician Roger Penrose, who was revolutionizing the tools used to study the properties of space-time.

In general relativity, space-time can behave strangely. These behaviors are part of the theory and not signs of a fatal flaw. Space-time can have folds, tears, edges, holes, creases, and be multiply connected and topologically complex.[30] The general relativity "landscape" is wildly different from that of Newton's gravity, which is based on three-dimensional space that is simple and linear everywhere. General relativity includes the possibility of singularities.

There are only two kinds of space-time singularity in general relativity. A singularity might be caused by matter being compressed to reach infinite mass density (as in a black hole). Or it might arise when light rays come from a place of infinite curvature and energy density (as in the big bang). The analogy for the first kind would be a flat sheet of paper with a hole in it or an edge (the second kind has no obvious analogy). Any particle traveling along the sheet of paper simply disappears when it encounters the singularity. Hawking and Penrose aimed for a general treatment. They stripped away as many assumptions as possible and proved a celebrated series of singularity theorems to show that singularities are inevitable in general relativity. In other words, they're a feature, not a bug. Every black hole must have a mass singularity, and every expanding universe (like ours) must have started with an energy singularity. Hawking used the cosmology example for his thesis and it made him an instant rock star in the rarified world of theoretical physics.[31]

Hawking then turned his attention to black holes. Along with two colleagues, he proposed that, like all the other objects in the universe, black holes are subject to the laws of thermodynamics. By this time, the mid-1960s, a full solution in general relativity for a spinning black hole had been found, adding to Schwarzschild's earlier solution for a stationary black hole. In mathematics or physics, a solution is the set of values for variables that satisfy all the equations. It's a sign of how very hard it is to find exact solutions in relativity that only two have been found in 100 years!

One of Hawking's black hole "laws" was that the surface area of black holes always increases. When matter falls into a black hole the area of the event horizon grows, and when two black holes merge the area of the resulting event horizon is bigger than the sum of the individual event horizons. This led to a new debate, with a startling conclusion.

In 1967, John Wheeler suggested that black holes were very simple objects, which could be described by just their mass and their angular momentum.[32] With his knack for a catchy name he called this the "no hair" theorem, hair being a metaphor for the details that characterize most physical objects. Jacob Bekenstein, one of Wheeler's graduate students, attempted to combine Wheeler's theory with Hawking's understanding of black hole surface area. Bekenstein argued that the area of a black hole was a manifestation of its entropy. In popular usage, entropy means disorder. In physics, entropy measures the number of ways atoms or molecules in an object can be rearranged without changing its overall properties. The "no hair" theorem implied that black holes don't have entropy, but Bekenstein pointed out that nothing observed in nature was immune from the second law of thermodynamics—entropy is always increasing—and black holes should be no exception.[33] As thermodynamics is a foundation stone of physics, Hawking accepted Bekenstein's argument, but he was then faced with a puzzle. If a black hole has entropy, it must also have a temperature. If it has a temperature, it must radiate energy. But if nothing escapes from a black hole, how can it radiate energy?

Hawking's answer to this conundrum stunned the theoretical physics community. He said that black holes evaporate. Here's how it works. In classical physics, the vacuum of space is empty. But in quantum theory, "virtual particles" are constantly being created and destroyed. They exist for tiny instants of time, allowed by Heisenberg's uncertainty principle. Normally these pairs of particle and antiparticle, or pairs of photons, disappear with no effect. But close to the event horizon of a black hole, intense gravity can pull the virtual pairs apart. One falls in and the other flies away and becomes real (Figure 8). This is how the black hole radiates energy. The energy needed to create the real particle comes from the gravitational field of the black hole, resulting in a decrease in mass. Riffing off Einstein's famous quip about quantum mechanics that "God does not play dice with the universe," Hawking declared that "God not only plays dice but also sometimes throws them where they cannot be seen."[34]

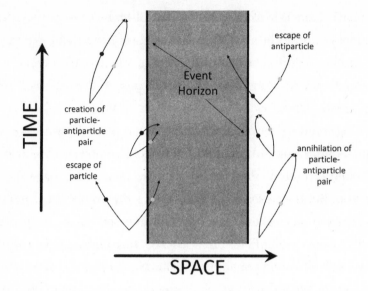

FIGURE 8. Black holes are not completely black. Virtual particle–antiparticle pairs are continually coming into existence and annihilating after a short period of time. According to a theory of Stephen Hawking, when this process occurs near the event horizon of a black hole, one member of the pair can escape while the other is captured by the black hole. The effect is that black holes radiate energy and will slowly evaporate. *Chris Impey*

Hawking radiation was controversial but undeniably brilliant, and Hawking was soon elected a Fellow of the Royal Society. Unfortunately, the effects of Hawking radiation are extremely small for a star remnant the mass of the Sun—a mere ten-millionth of a Kelvin, far too small for an astronomical measurement. The evaporation rate is amazingly slow. It would take 10^{66} years for a black hole the mass of the Sun to disappear completely. But the culmination of the process isn't dull. The temperature and the evaporation rate increase with decreasing mass, so black holes disappear in an explosive crescendo of radiation.

Black holes seemed to be getting stranger and stranger. Physicists explored their implications even as they doubted their existence. In 1935, Albert Einstein and Nathan Rosen had proposed the existence of "bridges" connecting two different points in space-time.[35] A black hole might be at either end of this bridge, which John Wheeler dubbed a wormhole. General relativity also allowed for regions of space-time that can't be entered from the outside, and yet allow light and matter to escape. These are known as white holes. A black hole region of the future might have a white hole region in its past. Wormholes and white holes have not been observed, but as Steven Weinberg once said, "This is often the way it is in physics—our mistake is not that we take our theories too seriously, but that we do not take them seriously enough."[36]

In popular culture, black holes became metaphors for death and destruction. But they also held out the hope for transformation and eternal life, since time freezes at the event horizon and nobody knows what lies inside. As novelist Martin Amis wrote, "Hawking understood black holes because he could *stare* at them. Black holes mean oblivion. Mean death. And Hawking has been staring at death his entire adult life."[37]

Betting on Black Holes

Stephen Hawking was a good person to bet with. His wagers rarely succeeded.[38] His first was on the cosmic censorship conjecture. In 1969, Roger Penrose proposed that singularities are always "hidden" behind an event horizon. With the exception of the big bang, there are no naked singularities. An event horizon prevents any observer from seeing matter get crushed to infinite density. Because a singularity presents big conceptual challenges for general relativity, physicists hoped that black holes always had event horizons. In 1991, Hawking bet two Caltech theorists, John Preskill and Kip Thorne, $100 that the cosmic censorship is correct and naked singularities don't exist. In 1997, supercomputer simulations showed that, under certain conditions, a collapsing black hole might lead to a naked singularity caused either by nature or perhaps by an advanced civilization. Hawking conceded, paid up, and gave his two colleagues T-shirts reading "Nature Abhors a Singularity."

That same year Hawking bet Preskill that information was destroyed in a black hole (Thorne switched sides and joined him in the bet). "Information" in this context is related to entropy. High entropy means disorder and a small amount of information. Normal gas, for example, is highly disordered and requires just a few bits of information to describe it: density, temperature, and chemical composition. Black holes have enormous entropy, much more than the balls of gas that form them, and accordingly they are described with even less information than gas: all we know is their mass and spin.[39] However, one could, in principle, make a black hole in an enormous number of different ways, such as by crushing together gas or rocks, or maybe books and mismatched socks— but you cannot see that information from outside. Then the black hole evaporates with the release of disordered radiation. What happens to all the information about what made the black hole in the first place? This conundrum became known as the information paradox.

In 2004, Hawking conceded this bet as well. At a conference in Dublin he reversed his earlier position and said that information could survive passage into a black hole, albeit in mangled form. It would be like burning an encyclopedia, and finding in the ashes and smoke feeble remnants of the information in it. Perhaps a clever calculation could reconstruct the patterns of ink and the text. Hawking preserved the tenets of quantum mechanics but he killed off an earlier speculation that information might not only be preserved inside a black hole, but might also pass into other universes branching off from a black hole. He told the *New York Times*, "I'm sorry to disappoint science fiction fans, but if information is preserved, there's no possibility of using black holes to travel to other universes."[40] Hawking was alluding to the idea in cosmology where the precursor state to the big bang could have spawned a multitude of universes, adding the notion that black holes might enable information to flow between the universes. Making good on his wager, Hawking gave his friend Preskill an encyclopedia of baseball, from which "information can be recovered with ease," and he called his initial claim of information loss his "biggest blunder."[41]

I met Stephen Hawking briefly when I was a graduate student in the late 1970s. He was giving a talk on black holes in London in honor of his appointment as Lucasian Chair of Mathematics. Hawking was thirty-six and at the peak of his powers as a physicist. He'd been in a wheelchair for a decade and his speech had deteriorated to the point where only a few family members and close colleagues could understand him. One of his students put his head close to Hawking's to be able to hear each phrase, then conveyed it to the audience. At the end of the talk, I remember having the powerful feeling that whatever obstacles I might come up against in my life or career, they would pale into insignificance compared to what Hawking had faced.

Twenty years later, my cousin and I went to see him give a public talk in a large auditorium in Cambridge. The lecture had been prepared in advance and it was delivered by the speech synthesizer that became his trademark. Questions went slowly because he had to use one finger to pick out phrases from a bank of thousands stored on the computer. His puckish sense of humor was on full display. Someone asked, "Will we ever be able to use black holes to save humanity from destruction?" Hawking paused, then tapped on the keyboard: "I hope not." Another question: "Could someone survive falling into a black hole?" Slowly, he tapped his response: "You might. I have enough to deal with already."

The true answer to this second question is that an unfortunate voyager falling into a black hole would not survive, as the stretching force of gravity would "spaghettify" them. Gravity falls off with the square of the distance from an object. For any compact object like a black hole, the difference in the gravity force between two points at different distances from the object can be large—this is a tidal force.[42] At a distance of 3,000 kilometers, the stretching force will create an acceleration between your head and your toes about equal to the Earth's gravity. That's uncomfortable but survivable. At a distance of 1,000 kilometers, the stretching force is 50 times Earth's gravity, so your bones and internal organs would be pulled apart. At 300 kilometers—still far from the event horizon—the stretching force is 1,000 times Earth's gravity, and solid objects will be destroyed. Spaghettification isn't like a child's game, in which someone pulls your feet and another person pulls your arms, or even like the medieval torture of the rack. Space-time near a black hole is being distorted, so you would be stretched at the level of muscle fibers and cells and strands of DNA.

This creates a paradox. The event horizon is a point of no return, an information membrane: information can get in but not out. If you could dive into a black hole carrying a digital clock and somehow avoid being spaghettified, the clock would seem to keep normal time as you plunged in free fall across the event horizon. Meanwhile, a companion watching you fall in would see your clock slow down while your distorted image slowly approached the event horizon, until both you and the clock appeared to stop. Now imagine we toss a book into the black hole. Gravity says it will cross the event horizon and the information will be lost. But to an outsider's perspective the book never reaches the event horizon. Is the information lost or is it "stored" somehow on the event horizon?

There's one bet that Hawking was happy that he lost: his first bet with Kip Thorne in 1975. Hawking bet against the existence of black holes as an insurance policy. He hoped to lose, but if he won he said he could console himself with a four-year subscription to the British satirical magazine *Private Eye*. As we'll see in the next chapter, the high-energy source Cygnus X-1 eventually proved to be a compelling black hole candidate, so Hawking conceded the bet in 1990. Kip Thorne's prize was a year's subscription to *Penthouse*.[43]

The Golden Age of Black Hole Theory

After Hawking's landmark findings, the pace of black hole research has accelerated. We're currently in a golden age of black hole theory, with a blizzard of papers published every year. Physicists are trying to reconcile the "smooth" descriptions of objects in general relativity with "grainy" descriptions of matter in quantum theory.

One of the biggest conundrums, as previously mentioned, is the problem of what happens to information at the event horizon. Hawking's theory of black hole evaporation reached into the toolkit of quantum mechanics. He originally argued that the radiation from a black hole is chaotic and random, and when the black hole evaporates all the information contained within it is lost. This violates a core premise of quantum theory, which says that particle interactions are time-reversible, so it should be possible to run the movie backward and recover the initial state from the final state. This collision between two highly successful theories of physics—general relativity and quantum mechanics—was considered to be a crisis by most physicists.

In 1996, Andy Strominger and Cumrun Vafa used string theory to reproduce Hawking's entropy and radiation.[44] String theory is a decades-long attempt to unify the four forces of nature with a conception of matter not as particles but as tiny one-dimensional "strings" of energy existing in a space-time that might have eight or ten dimensions. String theory is more fundamental than standard quantum theory because it postulates a single entity that underlies diverse particles such as electrons, protons, and neutrons. It's also appealing because it's mathematically elegant, but it has been difficult to test. Still, it was exciting when the theory was shown to explain some important properties of black holes, because this was the first success for a microscopic theory of matter in the realm of strong gravity. Strominger and Vafa's research suggested that information really could be recovered from a black hole. However, there is no clear agreement on how the information is preserved, or on what string theory can tell us about the nature of black holes.

Plenty of top-tier physicists are working on this puzzle.[45] One intriguing idea suggests that information is stored on the event horizon, in the way that a hologram is 2D information storage of a 3D object. If information about the contents of a black hole were somehow coded onto the surface (Figure 9), this would resolve the information paradox. In 2012, a big fly was found in the ointment: the virtual particles responsible for Hawking radiation are entangled, sharing quantum states even when they're widely separated. Getting information out by breaking entanglement would release a torrent of radiation, creating a "firewall" just above the event horizon. Rather than an uneventful journey into the dark abyss, a voyager would be obliterated by the firewall. But as seen from the outside, the voyager would still be trapped on the event horizon like a bug on flypaper. Do they die or survive? Nothing can get out, yet nothing can get in. Researchers are still arguing whether firewalls are inevitable or not.

This discussion illustrates the ebb and flow of ideas at the cutting

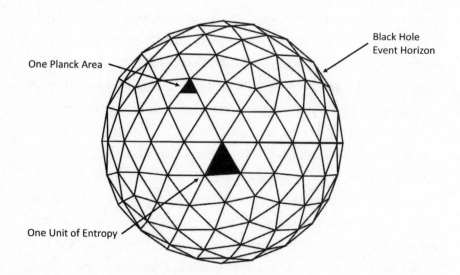

FIGURE 9. The entropy of a black hole is proportional to the area of the event horizon. Entropy can also be considered as information. The minimum unit or "bit" of information corresponds to a Planck area and it is determined by the speed of light, the strength of gravity, and Planck's constant. It is as if the contents of the black hole are written on the event horizon as bits of information. *Chris Impey*

edge of black hole theory. Let's leave the last word on the subject to Andy Strominger. In a 2016 paper called "Soft Hair on Black Holes," coauthored with Hawking, he argues against John Wheeler's "no hair" theorem and identifies particles that may act as the quantum pixels for information storage at the boundary of a black hole. It's still a work in progress. He admits, "I've got a list of thirty-five problems on the board, each of which will take several months. It's a very nice stage to be in if you're a theoretical physicist, because there are things we don't understand, but there are calculations that we can do that will definitely shed light on it."[46]

Over the past 100 years, black holes have evolved from a monstrous idea, one that violates common sense, to a proving ground for the most cherished theories in physics. Black holes are like a gift from the universe. They have heft, but the contents of the box are hidden and enigmatic. Yet even the wrapping is fascinating to study. I am reminded of Mark Twain's sardonic remark: "There's something fascinating about science. One gets such wholesale returns of conjecture out of such a trifling investment of fact."

It's time to ask a pragmatic question: do black holes actually exist?

2.

BLACK HOLES FROM STAR DEATH

SCIENCE HINGES ON the interplay between theory and observation. Over the millennia, humans have had many imaginative ideas about how the universe works. But without data drawn from observations, even the cleverest idea remains in the realm of speculation. Is there actually any evidence that mass can disappear from view in the universe?

Despite the challenge of imagining them, black holes are real. That's the firm conclusion from nearly fifty years of research on the end states of stars. An isolated black hole is completely invisible. The rupture it creates in space-time is so small that it's undetectable with any telescope. But most stars are in binary or multiple systems, so the visible star can be a pointer to the dark companion.

The Forces of Light and Darkness

When you look at the Sun, it's hard to believe you're watching a titanic battle between the forces of light and darkness. Although the Sun seems barely to change from day to day or year to year, particles are careering around at near the speed of light and planet-size parcels of plasma are constantly churning; it's a thermostatically controlled nuclear furnace.

At every point within it there is a balance between the inward force of gravity and the outward force from radiation released in the fusion of hydrogen into helium.[1] As long as the fuel for fusion remains, neither force gains the upper hand.

If you're placing a bet on the long-term outcome of this battle, gravity would be the smart choice. Nuclear fuel is finite but gravity is eternal. After hydrogen is exhausted in stars like the Sun, the interior pressure is lost and the core of the star collapses to a hotter and denser configuration where helium can be fused into carbon. This reaction goes quickly, and when the helium is exhausted the temperature can't rise high enough to ignite new nuclear reactions. The pressure support is lost and the core of the star is once again faced with gravitational collapse. The Sun will have a brief pyrotechnic phase as the last fuel is used up, ejecting about a third of its mass in a shell of gas moving at supersonic speed. The fast-moving gas heats up and glows, producing the gorgeous hues of a planetary nebula. Anyone watching the Sun from another star system in 5 billion years will see a spectacular light show. Anyone watching from the Earth will be in a lot of trouble, since the ejected gas will vaporize the biosphere and obliterate all life.

A star's life and death is governed by its mass (Figure 10). The diverse fates of stars are all preordained at birth. Depending on their masses, all stars will become either white dwarfs, neutron stars, or black holes. There isn't a "typical" mass or size for a star, although the process by which stars form from chaotic clouds of gas produces more small stars than large stars by a substantial factor. The Sun is toward the lower end of the mass range, and below it are dim stars called red dwarfs. There are several hundred times as many red dwarfs as there are stars like the Sun. Lifetime is also dictated by mass, since gravity determines the temperature of the core, which in turn indicates how quickly nuclear reactions will run and therefore how long the nuclear fuel will last. A star like the Sun will fuse hydrogen into helium for 10 billion years; we're halfway through that span.[2] A star half the mass of the Sun has a

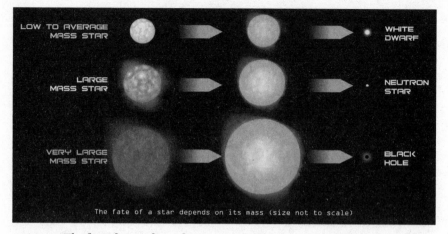

LOW TO AVERAGE
MASS STAR

WHITE
DWARF

LARGE
MASS STAR

NEUTRON
STAR

VERY LARGE
MASS STAR

BLACK
HOLE

The fate of a star depends on its mass (size not to scale)

FIGURE 10. The fate of a star depends on its mass. Most stars, including the Sun, are low to average mass and after their nuclear fuel is exhausted they will die as cooling embers called white dwarfs. More massive stars have more fuel but shorter lives and they die as neutron stars or black holes. *NASA/Chandra Science Center*

lifetime of 55 billion years, so no star of that mass has ever died in the history of the universe, which has only spanned 14 billion years. A red dwarf a tenth the mass of the Sun, which is as puny as a star can be and still have fusion reactions, will spend its fuel like a miser. Such a star would theoretically live more than a trillion years—an unimaginably long time. Even so, the dwarf star is just delaying the inevitable because one day the fuel must be exhausted, the dim light must flicker out, and gravity will be rewarded for its patience.

Stars more massive than the Sun have shorter and more spectacular lives. They all do what the Sun is doing now—fuse hydrogen into helium—but they have more gravity so have hotter cores and use up their fuel at a ferocious rate. The more massive the star, the hotter its core temperature and the shorter its lifetime. Massive stars can fuse all the elements in the periodic table up to iron, the most stable element. When nuclear reactions stop at iron, the core is in a bizarre physical state: it's an iron plasma 100 times denser than water at a temperature of a billion degrees. Without pressure from the core, it collapses, and the compression wave inward bounces into a multi-billion-degree blast wave outward, wherein heavy elements up to uranium are fused in split seconds. This is a supernova, one of the most dramatic events in the universe. Precious metals are flung into space to become part of a next generation of stars and planets. Much of the original mass of the star is ejected, but what remains is squeezed tightly by gravity's unrelenting grip.

Gravity and Darkness Are the Final Victors

The remnants of stars are truly bizarre states of matter. We have no way to create them in the lab. All we can do is use the laws of physics and hope that our theories are sturdy enough for the task. Some of the best minds in twentieth-century astrophysics were consumed with understanding stellar remnants.

A star's aftermath depends on the mass a star has when it starts its life. Stars are born in the fragmentation and collapse of large gas clouds that produce many more low-mass stars than high-mass stars. All stars lose some fraction of their mass as they age. The processes by which this happens are complex, so boundaries between the different outcomes are not precise. Stars that start their lives below 8 times the mass of the Sun collapse to an unusually dense state of matter called a white dwarf. The vast majority of stars are less massive than the Sun, so over 95% of all stars will end up this way. For example, the Sun will shed about half its mass during its pyrotechnic late stage of life before dying as a white dwarf.

The English astronomer William Herschel accidentally discovered a star called 40 Eridani B in 1783, but he had no way to measure its size so he didn't realize that the star was unusual. In 1910, astronomers refocused their attention on this dim star, which is in a binary system. The orbit revealed its mass to be about the same as that of the Sun. They knew its distance, and deduced that it was 10,000 times fainter than the Sun would be at the same distance. Yet it was white, therefore much hotter than the Sun. To see why this is puzzling, think about electric hot plates on a stove, viewed in a darkened room. One hot plate is turned on low and glows orange, like the Sun. A second hot plate is turned on high and is much hotter, so it glows white. The white hot plate is much brighter than the orange hot plate. For the white hot plate to appear much fainter than the orange hot plate, it would have to be much smaller. By the same logic, the faint star in the 40 Eridani system had to be much smaller than the Sun. With the same mass as the Sun, it had to be much denser as well.[3]

Ernst Öpik calculated that 40 Eridani B should have a density 25,000 times higher than that of the Sun, which he called "impossible."[4] Arthur Eddington, who popularized the term "white dwarf," described the incredulous reaction a white dwarf produces: "We learn about the stars by receiving and interpreting the messages which their light brings to us. The message . . . when it was decoded ran: 'I am composed of material 3,000 times denser than anything you have ever come across; a ton of my material would be a little nugget that you could put in a matchbox.' What reply can one make to such a message? The reply most of us made in 1914 was—'Shut up. Don't talk nonsense.'"[5]

Eddington was not a humble man. When a colleague said to him, "Professor Eddington, you must be one of only three people in the world who understand relativity," he paused, so the colleague said, "Don't be so modest." Eddington replied, "On the contrary, I'm trying to think who the third person is."[6] Even though Eddington was a master of the astrophysics that predicted white dwarfs, he called them "impossible stars."

A typical white dwarf is the size of the Earth but has the mass of the Sun. Its density is a million times higher than water. With no energy release from fusion, and so no outward pressure, gravity shrinks the gas, crushing the atomic structure and forming a plasma of unbound nuclei and electrons. Only at this point is gravity finally thwarted. In 1925, Wolfgang Pauli came up with the exclusion principle, which says that no two electrons can have exactly the same set of quantum properties. Its effect is to provide pressure that stops the stellar corpse from collapsing any further.[7] A white dwarf will form with a temperature as hot as 100,000 Kelvin, and then steadily radiate its heat into space. Fade to black.

Subrahmanyan Chandrasekhar, at the time a nineteen-year-old Cambridge student on an Indian government scholarship, calculated that regardless of the starting mass of a star, its white dwarf remnant can never be larger than about 1.4 times the mass of the Sun. Above this mass, gravity trumps quantum mechanics and the star collapses to a singularity. The maximum mass of a white dwarf is called the Chandrasekhar limit.[8] It was a brilliant calculation—so Chandrasekhar was understandably disappointed when Arthur Eddington, his idol, publicly ridiculed the idea of collapse to a singularity. Chandrasekhar felt betrayed, believing that the slight was in part racially motivated. We'd like to think science is a meritocracy, but scientists can be jealous and short-sighted. (Quantum pioneer Paul Dirac, who experienced similar resistance, pithily observed that science advances one funeral at a time.) Chandrasekhar was eventually vindicated, and won the Nobel Prize in Physics for his insights into the structure and evolution of stars.

Chandrasekhar opened the door for physicists to imagine what happens if a star collapses beyond a white dwarf. A few years later, California astronomers Walter Baade and Fritz Zwicky suggested, almost casually, that above the Chandrasekhar limit, pure neutron material might result from star collapse, but they didn't do any calculations to support the conjecture. In 1939, chain-smoking, hard-driving Robert Oppenheimer did the math. With a graduate student, he established the mass range of neutron stars.[9] The same year, as we've already seen, he showed that with a stellar remnant above this mass range—more than 3 times the mass of the Sun—a black hole must form.

All stars lose mass before they die. As mentioned above, the Sun will lose half of its mass before it dies as a white dwarf. All stars beginning their lives with up to 8 times the mass of the Sun will leave behind white dwarfs with masses up to 1.4 times the mass of the Sun. If the initial mass of a star is roughly between 8 and 25 times the mass of the Sun, the core collapse continues until all the protons and electrons merge into pure neutron material.[10] Since there's no electrical force, the neutrons jam together like eggs in an egg carton. The material is supported against further collapse by the strong nuclear force and a stronger version of the quantum force that prevents white dwarfs shrinking further. This is a neutron star, the smallest and densest type of star in the universe. Above 25 solar masses, we're faced with the prospect of Einstein's monster (Figure 11).

Neutron stars challenge the imagination.[11] A neutron star is like a city-sized atomic nucleus with an atomic number of 10^{57}. Its material is 1,000 trillion times denser than water. A sugar cube amount of white dwarf material brought to the Earth would weigh a ton, but a sugar cube

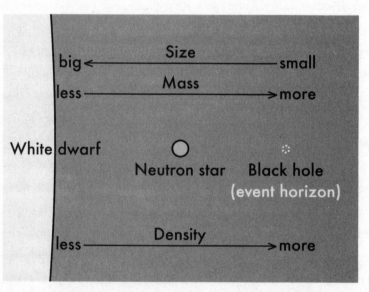

FIGURE 11. The greater the initial mass of the star, the smaller and denser the stellar remnant after all fusion is complete. The curve on the left shows the size of a white dwarf. Even though neutron stars and black holes are only a few times more massive than white dwarfs, the symbols show that they are much smaller and denser. *Patrick Len/Cuesta College*

amount of neutron star material brought to the Earth would weigh as much as Mount Everest. When a star shrinks this much, the magnetic field is squeezed and concentrated too. Some neutron stars have magnetic fields a quadrillion times stronger than the Earth's. The gravity near the surface is so strong that an object falling from a height of a meter would accelerate to 3 million miles per hour at the moment of impact. Conservation of angular momentum means the normally sedate rotation of a star like the Sun is amplified when a star collapses. The fastest spinning neutron star spins 716 times per second, or 42,000 rpm. Such a rapidly spinning solid object isn't completely stable, so the solid crust can violently shift in an event called a starquake.

How can a neutron star be detected? These city-sized stars should emit no light because they are not fusing elements the way normal stars do. For a couple of decades astronomers consigned them to the category of astrophysical curiosities: something to be imagined but never witnessed. Then in 1967, young graduate student Jocelyn Bell and her thesis advisor, Tony Hewish, detected radio pulses with a period of 1.3373 seconds from an unknown object in the constellation of Vulpecula. The pulses were so powerful and regular that Bell and Hewish thought the object might be a beacon, so they jokingly named it LGM-1 (for "Little Green Men"). Other "pulsars" were soon discovered, and Bell and Hewish made the connection with the earlier prediction of neutron stars. The intense magnetic field drives radio emission from hot spots on the neutron star's surface, and when the spinning neutron star sweeps that emission across a radio telescope, like a lighthouse beam, pulses are seen.

Controversy erupted seven years later when a Nobel Prize for the discovery of pulsars was awarded to Hewish and Martin Ryle, the head of the radio observatory, but not to Jocelyn Bell, who made the actual discovery. It is clear to many in the scientific community that she was excluded from the honor because she was a young woman. Just over 200 scientists have won the Nobel Prize in Physics, and only two have been women: Marie Curie in 1903 and Maria Goeppert-Mayer in 1963.[12]

Surveys with radio telescopes have steadily increased the number of pulsars to over 3,000. However, the conditions that lead to a hot spot are rare, so very few neutron stars are radio pulsars. The vast majority of the millions of neutron stars in the galaxy are spinning quietly in deep space, dark and undetectable.

Finding the First Black Swan

It was 1964. The Beatles took America by storm and a brash young fighter called Cassius Clay became heavyweight champion of the world. Science was booming as well. The term "black hole" appeared in print for the first time in January 1964, and in June a small sounding rocket launched from New Mexico identified a strong source of X-rays in the constellation of Cygnus, the swan. "Black swan" is the term for those rare, unexpected events that play a disproportionate role in the development of science. (It's also used by philosophers to talk about the problem of induction: seeing many white swans is not evidence that black swans don't exist.) Finding the first example of a black swan in black hole physics involved detective work that took seven years.[13]

X-ray astronomy was a new field in the 1960s. High-energy radiation from cosmic sources can only be detected from space; the first source had been found just two years earlier. The eight sources identified in the 1964 observations were consistent with supernova remnants, or hot gas created when a massive star dies violently.[14] The discovery observations had poor spatial resolution, so they didn't narrow down the location of the X-rays emanating from Cygnus to a region much smaller than the constellation itself. In 1970, the Uhuru X-ray satellite showed that the intensity of Cygnus X-1 varied in less than a second. Astrophysicists use time as a way to measure the size of remote objects; the idea is that intensity changes cannot occur any quicker than the time it takes light to cross the source of light. The intensity variations of Cygnus X-1 suggested that the object could be no larger than 100,000 kilometers across—less than a tenth the size of the Sun.

An accurate position on the sky from the National Radio Astronomy Observatory identified the variable X-ray source with a blue supergiant star called HDE 226868. Supergiants are hot stars but they're incapable of emitting copious amounts of X-rays. The only explanation for the X-rays was that something in that region of space was heating gas to a temperature of millions of degrees. The decisive next step used optical techniques. In 1971, two groups of scientists took spectra of the blue supergiant and found periodic variations in the Doppler shift of the star that matched the variations in the X-ray emission.[15] Orbital calculations allowed the researchers to estimate the mass of the "invisible" companion that was tugging at the supergiant. The researchers speculated that a black hole was sucking gas off a companion star and that this gas was somehow heated to a temperature high enough to produce X-rays (Figure 12).

As astronomer Tom Bolton prepared to present a paper on these findings at the American Astronomical Society meeting in Puerto Rico, he was wracked with nerves. He was only twenty-eight years old. "Five minutes before I gave my paper I was revising it on the fly. I was sitting at the back of the room trying to get the latest data in my diagram," he recalled.[16] He was also feeling the pressure of competition. He was just one year out of his PhD and working alone. A more experienced team at the Royal Greenwich Observatory, using a larger telescope, was getting similar data on Cygnus X-1. Everyone was cautious about their interpretations, since careers had previously foundered on false claims of having detected a black hole. Within a year, Bolton was sure, and he staked his reputation on it. He next presented the research at the Institute for Advanced Study in Princeton—the academic home of Einstein and Oppenheimer. The observations were robust. The audience was convinced. The first black swan had been found.

By the late 1970s, black holes had entered popular culture. Their

FIGURE 12. The prototypical black hole is the strongest X-ray source in the constellation of Cygnus, Cygnus X-1. The black hole is in a tight binary orbit with a blue supergiant star. Gas pulled toward the black hole forms an accretion disk, which heats up so much that it emits copious X-rays. *NASA/Chabdra Science Center Artist/M. Weiss*

bizarre properties fascinated people who rarely gave a thought to astronomy. Disney released a movie called *The Black Hole,* and its ominous theme earned it a PG rating, a first for Disney. Although the movie is low-tech and cheesy in parts, it was ambitious for its time and it presented black holes as metaphors for death and transfiguration. The pop innocence of the Beatles evolved into raucous rock. Rush, Queen, and Pink Floyd all gave nods to astrophysics.[17]

Weighing the Invisible Dance Partner

For any star, mass is destiny. Mass indicates the size of the star's fuel tank for fusion reactions. Mass also determines the star's gravity, so it dictates the size of a star, its internal temperature and pressure, what kind of fusion it can support, and how fast nuclear reactions occur: all this from one number. Any claim of black hole detection must be grounded in a reliable estimate of mass. Unfortunately, mass is also the hardest quantity to measure. Visual data give brightness and surface temperature, but separate observations are required to measure distance and therefore luminosity, and then a stellar model is needed to infer mass.

A solo black hole lurking in deep space has enormous mass, but it is undetectable on its own. Luckily, over half of all stars are in binary or multiple systems. Newton's law of gravity says that two objects tug on each other with equal force. They orbit a common point called the center of mass and always stay on opposite sides of it. Imagine two people holding hands and spinning around. If they are the same weight they will "orbit" a point that's midway between them. But if an adult spins a child they'll pirouette around a point closer to the adult than to the child, more like a hammer throw (and I hope the analogy ends there). It works the same way for stars. Two stars with equal mass orbit at equal distances from the center of mass. If the two masses are unequal, the massive star is closer to the center of mass while the less massive star has a greater acceleration and travels faster in a larger orbit (Figure 13).[18]

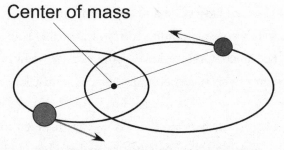

Center of mass

FIGURE 13. In a binary star system, the two stars orbit a common center of mass. The more massive star is closer to the center of mass and the less massive star is farther from the center of mass. If the stars are too close to resolve in an image, the orbit can be measured with spectroscopy, where spectral lines shift to redder and then bluer wavelengths in a cycle that gives the period of the orbit. *Robert H. Gowdy/Virginia Commonwealth University*

That's the concept. Now let's add the math. In a circular orbit, the velocity is the circumference divided by the time taken to complete the orbit, or the period. If we measure the period and the velocity, we can get the radius of the orbit. Newton's form of Kepler's third law of motion relates the combined mass of two stars in an orbit to the size and the period of the orbit. That's four variables, so we need to measure three of them. Therefore, in a binary system with one visible star and one invisible companion, we must measure the mass of the visible star in order to nail down the mass of the dark object.[19] How do we do that?

The dance floor is dark. The woman is dressed in white. The man is dressed in black. With dim lighting from the side, the woman can be seen but the man is invisible. They twirl across the floor. By the way she moves, we know the woman is in the grip of an unseen companion. Binary stars are in a similarly tight embrace and oblivious to the larger universe. If the pair of stars is widely separated and not too far from the Earth, we can see both stars and simply watch their motion to measure the orbit. This is a visual binary. More often, the stars are far away and astronomers can't see them as separate objects, but spectroscopy reveals absorption lines from each star that oscillate to longer and shorter wavelengths, representing a periodic Doppler shift caused by the orbital motion. This is a spectroscopic binary. In a binary where one star is a black hole, we have one hand tied behind our backs because the spectrum only shows absorption lines from the visible star.

As with the dancers, the motion of the star that can be seen indicates the motion of the star that can't be seen. But there are two big complications. The first is that we need to estimate the mass of the visible star. To do this, we need to determine the distance to the binary system so as to calculate the luminosity, or the number of photons emitted by the star every second. Then these quantities, along with the surface temperature of the star (determined from its color) and its surface gravity (determined from the shape of lines in the spectrum), are fed into a sophisticated model of stellar structure and energy production to generate an estimated mass.

Second, there's the problem of our perspective. Spectroscopy measures a Doppler shift, or radial motion toward or away from the observer. A binary system seen on its edge—where the orbit is perpendicular to the plane of the sky—gives the full effect, since once every orbit each star is coming directly toward us and the other star is going directly away from us. But if a binary system is seen face-on—where the orbit is in the plane of the sky—no Doppler effect is measured, since all the motion is side-to-side. Binaries are scattered at random orientations in space, so we're faced with an additional complication in that we don't know the inclination angle. The good news is that for almost all inclination angles the Doppler shift underestimates the orbital velocity, since usually some part of the motion is not radial. So when astronomers calculate a stellar mass, they generally can only determine its lower limit. Since the goal is to prove that the unseen companion has the minimum mass required for it to be a black hole, the method works.[20]

Black Holes with Gold-Plated Credentials

When people think about astronomy, they think of the gorgeous images taken by the Hubble Space Telescope. But many advances in understanding the universe have come from spectroscopy, the technique of dispersing light into its constituent colors. Newton used a spectrum to understand the nature of light. In the early 1800s, young Joseph von Fraunhofer survived an orphanage, a brutal overseer, and an explosion in the glassmaking factory where he worked, and went on to make the first spectrum of the Sun and see features that hinted at its composition. A hundred years later, a group of poorly paid women at Harvard College Observatory scanned hundreds of thousands of spectra recorded on photographic plates, gathering information that was used to understand what stars are made of and the true size of the universe.[21]

I've taken thousands of spectra in my life as an astronomer, and each one is a puzzle to solve or a present to unwrap. They're key to measuring distance and chemical composition, and they yield clues to the unspeakable violence in the centers of galaxies. That squiggle on the screen at the end of a night of observation was the result of light arriving at the telescope, being smeared into a feeble streak by the spectrograph, and falling onto the silicon CCD, or charge-coupled device. A CCD converts photons into electrons and then into an electrical signal that is processed into a map of intensity versus wavelength.

One night in Hawaii, I was observing at a telescope atop the 14,000-foot dormant volcano Mauna Kea. Data from the CCD was arrayed as horizontal streaks on the computer monitor. My eye was drawn to one particular faint streak. The dark notches in the digital spectrum were evidence of a distant galaxy made of the same elements as the Milky Way. I could infer its rotation and the kind of stars it was made of and the amount of gas mixed in among the stars. The redshift of the spectral features told me the galaxy was 10 billion light years away and the light had been traveling since long before the Earth formed. I knew that this faint galaxy had been moving away from the Milky Way faster than the speed of light when the light was emitted, thanks to the extremely fast expansion of the universe soon after the big bang. The universe is governed by general relativity and not special relativity, so space can expand faster than light speed! I'm slightly ashamed to admit it, but at the time I even forgot to be amazed that I could know such things about the universe. I rarely questioned the chain of reasoning and the foundation of scientific method that underpinned what I knew.

Spectroscopy is the key to understanding binary stars and their orbits. It lets astronomers measure the mass of the unseen companion in a binary system. And it lets them measure the unseen mass with enough precision to conclude that Einstein's monsters are real. There are a modest number of "gold-plated" cases of a binary system where the unseen companion has mass sufficient to be a black hole, and they're extremely difficult to explain with any other hypothesis. Let's take a closer look at the archetype, Cygnus X-1.

From the Earth, we look toward the constellation of Cygnus, riding high in the summer sky. We home in on a region near the center of the cross marking the body of the swan. With good binoculars we can see a blue-white star nestled in a loose group of hot, young stars that all formed at the same time. Five million years ago, when our primate ancestors split off into a branch of the evolutionary tree, these stars congealed from a collapsing cloud of gas and dust. The blue-white star of interest is 6,000 light years away, near the edge of a neighboring spiral arm in the Milky Way. That's a prodigious distance of 20,000 trillion miles. For it to be visible so easily the star must be extremely luminous, emitting 400,000 times more energy than the Sun. The light is old. It left the star when there were less than a million humans on the planet and mammoths were about to go extinct in North America.

We approach our prey cautiously. At the distance of the Earth from the Sun, the star would be blindingly bright, 20 times the size of the Sun—the width of two spread hands held at arm's length. This blue supergiant is locked in a six-day orbit with a near-invisible companion closer to it than Mercury is to the Sun. But the companion isn't totally dark. The blue supergiant is a ferocious fusion reactor and it pushes a wind of plasma from its outer atmosphere into space. Some of that material is pulled in by the companion to form a swirling disk of super-hot gas. At a temperature of over a million degrees, the gas disk emits copious ultraviolet radiation and X-rays. The companion's gravity also distorts the outer envelope of the supergiant into a teardrop shape with its narrow end pointing toward the companion. If we could follow the pointing teardrop and approach the swirling disk that marks the companion, we would see a small jot of utter darkness at its center: the black hole (Figure 14).

FIGURE 14. A completely isolated black hole would be undetectable. In a binary system with a massive star, the black hole siphons mass from its companion onto an accretion disk, and that gas is heated enough to emit X-rays. The accretion disk swirls around the event horizon of the spinning black hole. *NASA/JPL-Caltech.*

This description is inference; we've never seen this or any other black hole up close. Still, over 100 research papers have been written on Cygnus X-1; it's one of the most intensively studied objects in the sky. The orbital period has been measured with exquisite accuracy: 5.599829 days with an error of a tenth of a second.[22] We need to know the supergiant's mass and the inclination of the orbit to calculate the mass of its companion. Spectroscopy and detailed modeling shows that HDE 226868 is roughly 40 times the mass of the Sun.[23] Measuring the inclination is tougher, since the dark companion never goes behind the visible star; in other words, the system doesn't show eclipses. Recent work indicates an inclination of 27 degrees, which implies that the mass of the dark companion is 15 times the mass of the Sun.[24] That's far beyond the maximum stellar remnant mass for a neutron star; its gravity is so strong that the compact companion must be a black hole. All of the uncertainties in the data and all of the uncertainties in the modeling do not soften this conclusion.[25] By 1990, the evidence was good enough for Stephen Hawking to slip into Kip Thorne's Caltech office and sign a certificate on the wall, conceding their bet.

Stars massive enough to die as black holes are very rare. The Milky Way galaxy contains about 400 billion stars, most of which are dim red dwarfs far less massive than the Sun. We can use the small sample of confirmed black holes in the neighborhood of the Sun to project the total population in the entire galaxy, which yields an estimate of 300 million black holes. The few dozen gold-plated examples are a near infinitesimally small fraction of the total population, which is in turn a tiny fraction of all stars.

In the past decade or so, experts have published lists of 25–30 gold-plated black hole candidates.[26] The number increases slowly because of the high bar set on the evidence. All are in binary systems with extremely well-measured orbits, where the dark companions have masses more than 3 times that of the Sun and so must be black holes. In each case, there are additional strands of evidence that support the hypothesis. The masses of these black holes range from 6 to 20 times that of the Sun, while their orbital periods range from a leisurely month to a speedy four hours. Two black holes were found in the nearest neighboring galaxy to the Milky Way, the Large Magellanic Cloud: LMC X-1 and LMC X-3, both at a distance of 165,000 light years. All the others range from 4,000 to 40,000 light years away from the Earth. Another 30 systems are waiting for more or better data to join the gold-plated list.

Using Gravitational Optics

So far in our story, the search for black holes has depended on binary star systems, where the black hole is the invisible dancer. However, there's one method that can find the dark dancer even if it is alone. This method is based on a central prediction of general relativity: the deflection of light by any mass. Since mass bends light, a star or a galaxy can focus and magnify light from a more distant source. This phenomenon is called gravitational lensing, and was predicted soon after Einstein published his theory. It wasn't actually observed until 1979, when two images of a single quasar were observed; the splitting was caused by an intervening cluster of galaxies.

Lensing is a subtle effect; a single star isn't massive enough to bend light much. In 1919, Eddington measured the deflection of light from a distant star passing by the edge of the Sun at two arc seconds—a thousandth of the Sun's angular diameter. Lensing is also rare; the space between stars is vast, and it's unlikely that any two will line up closely enough for lensing to be observed. The odds of such an alignment are one in a million, so a million stars might have to be watched to catch one event. When a nearby star passes directly in front of a more distant star, the effect is called microlensing. With microlensing, the deflection angle is too small for image splitting to be seen, but there's gravitational amplification of the background star's light. An observer sees the background star temporarily brighten as the foreground star crosses it. The heavier the foreground star, the longer the duration of the effect. Since lensing depends on mass not light, the temporary brightening occurs even when the foreground star or the lens emits no light (Figure 15). This is the only way to detect an isolated black hole.[27]

An upside of microlensing is that the method is simple and direct. With any binary star system there are two masses to measure, an orbital inclination that's often unknown, and parameters that are derived indirectly from spectroscopy. Lensing involves a single equation that relates amplification to the mass and distance of the lens. For typical black hole masses the amplification lasts hundreds of days, so it's easy to observe. A downside of microlensing is that amplification is a one-time event, unlike the repetitive orbit of a binary system, which allows more data to

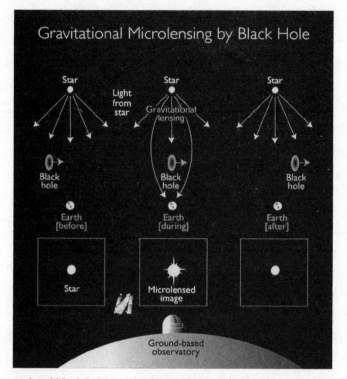

FIGURE 15. Isolated black holes can be detected using the fact that mass bends light. If a black hole passes directly in front of a more distant star, the black hole acts as a lens and the light of the star is briefly amplified. The image splitting is too small to be seen by any telescope. *NASA/ESA*

be gathered in the future. When a black hole passes in front of a more distant star, they're like ships passing in the night. The signal never repeats. More important, distance and mass are related in the lensing equation—so unless extra information is available to nail down the distance, the mass is uncertain.

Hunting black holes by microlensing is a needle-in-a-haystack activity. Microlensing surveys were developed to look for MACHOs, or massive compact halo objects that might explain the "dark matter" which outweighs normal matter in our galaxy by a factor of 6. MACHOs can be any kind of object that is dark or very dim, like black holes, neutron stars, brown dwarfs (sub-stellar objects), or free-floating planets.

Microlensing was unsuccessful in detecting MACHOs, but these same surveys meant to detect dark matter did detect (a few) black holes.[28] One in a million stars will experience microlensing, but only 1% of those will be lensed by a black hole, so several hundred million stars must be monitored to find a couple of black holes. A Polish research group used a decade of data from a 1.3-meter telescope to sift three good black hole candidates from billions of photometric measurements for 150 million stars.[29] Now that's dedication.

Physics at the Edge of the Maelstrom

The narrator of Edgar Allan Poe's 1841 short story "A Descent into the Maelström" is a young man who ages suddenly as he anticipates his likely demise in a whirlpool off the coast of Norway. One of his brothers dies in the abyss; another is driven insane by the spectacle. The narrator alone survives to tell the tale.[30] He winces as he recalls the scene: "The edge of the whirl was represented by a broad belt of gleaming spray; but no particle of this slipped into the mouth of the terrific funnel, whose interior, as far as the eye could fathom it, was a smooth, shining, and jet-black wall of water. . . ."

Poe's fictional narrator found a strange and terrible beauty in the maelstrom. We might feel similarly about a black hole. Einstein's monsters are terrifying but riveting. Like the gleaming spray and flotsam and jetsam at the edge of the maelstrom, a black hole in a binary system gives rise to spectacular effects. It's a wonderful irony of astronomy that objects which are in principle totally invisible can be the brightest objects in the universe. The reason is gravity.

For a terrestrial example, consider the Itaipu Dam at the border between Brazil and Paraguay. This facility generates a phenomenal amount of power, 100 terawatt hours per year, enough to fulfill the energy needs of several hundred million people.[31] Where does this power come from? The dam creates an elevated body of water from the Parana River. Every second, 300,000 cubic meters of water falls 110 meters, converting gravitational potential energy to kinetic energy as it accelerates to 100 mph. The water is slowed down to 10 mph at the bottom of the dam as kinetic energy is turned into rotational energy by turbine blades; the spinning turbine generates electricity. In a similar way, matter falling into a black hole generates energy.

Let's look at what happens when matter falls into a black hole. This process is called accretion. Black holes mostly attract the hydrogen gas that forms stars and loosely fills the empty space between them. These protons and electrons *could* fall straight in. They could plunge toward the event horizon, disappear into the black hole, and never be seen again. However, that's very unlikely, because few gas particles will be heading directly toward the black hole; most of them will have a sideways motion. With some sideways motion, the particles might head into space, never to return, or they might start orbiting the black hole. They'll also collide with one another, since they have slightly different trajectories. So particles take an indirect, chaotic route toward the black hole, and with all the collisions the gas gets hot.

Now we add the crucial fact that the black hole spins, which gives it angular momentum. In physics, angular momentum is always conserved; there's a rule about how particles move in a rotating system.[32] The black hole spins quickly because it has collapsed and is small; a slowly spinning massive star will turn into a rapidly spinning stellar remnant (think of a skater spinning slowly when their arms are spread out, then faster as they bring their arms in to be more compact). The spinning black hole swirls the hot gas around it, just as water near the edge of a bathtub would swirl if you vigorously stirred the water at the center. The swirling of the gas is strongest around the black hole's equator. This vortex of hot gas is called an accretion disk.

Since most of the gas is concentrated in an accretion disk around the black hole's equator, the regions above the poles of the black hole are relatively empty. That means some of the hot gas can escape along the poles. As it does so, it converts spin energy from the black hole into kinetic energy. That gas is ejected in twin jets of fast-moving particles that are aligned with the spin axis of the black hole. Those jets carry away a small portion of the gravitational energy of the matter that's falling in. If we could get close to the accretion disk we would see bizarre distortions due to the black hole's intense gravity bending light (Figure 16).

Marck's enhanced image: black hole lit by accretion disc.

FIGURE 16. Image from a computer simulation of a black hole and surrounding accretion disk, using the full equations of general relativity. The brighter, left side of the disk is approaching and the darker, right side is receding. The distortions are caused by mass bending light. Note that the far side of the disk is not hidden by the black hole because gravitational lensing lets us see behind the black hole. *J. A. Marck/CNRS*

We can visualize the swirling disk of gas, just like the flotsam and jetsam at the edge of Poe's maelstrom. The center of the action is a spinning black hole, dark and implacable. As particles get closer to the black hole, they move faster. Their gravitational energy is being converted into kinetic energy. They're also colliding, so the gas heats up, and friction within the disk causes intense thermal radiation. Gas in an accretion disk has a temperature of millions of degrees and glows brightly in X-rays.

Thus, gravity power is turned into radiation. There's irony in the fact that something so black can create a scene that's so bright. The process is extremely efficient. Efficiency in this context refers to the fraction of stored energy that is converted into radiation. Chemical burning, the source of most of our power on Earth, has an efficiency of 0.0000001%. Stellar fusion, the process that makes stars shine, has an efficiency of just under 1%. Accretion onto a stationary black hole has an efficiency of 10%, increasing to 40% for a spinning black hole.[33] Black holes are the most powerful energy sources in nature.

Gas doesn't fall easily into the black hole because it has angular momentum. The same is true for planets orbiting a sun. Working out the details of black hole accretion was one of the most challenging problems in astrophysics; it took dozens of researchers nearly two decades to solve.[34] Gas particles in the disk experience friction, so the entire disk acts as if it's viscous. As a result, some material loses angular momentum and moves closer to the black hole while some material gains angular momentum and moves farther out. Particles that get close to the disk's inner edge are moving within a hairsbreadth of the speed of light. As it approaches the event horizon, a typical particle experiences a slow spiraling in through the accretion disk, jostling with all the other particles. Then, at the inner edge of the accretion disk, gravity pulls it directly into the black hole. A black hole gathers mass through this sequence of events.

The limit to accretion was calculated by Sir Arthur Eddington early in the twentieth century. The Eddington limit assumes a spherical geometry and asks at what point the gravitational force pulling a particle inward is matched by the radiation pressure pushing the particle outward. The maximum rate at which mass can be added to a black hole is rather low: in a year, it can grow by no more than a third of the Moon's mass. At that rate, it would take 30 million years to double in mass. But the efficient conversion of mass falling in to radiation going out means the black hole is blindingly bright. A black hole fueled by gas from a companion can be 100 times brighter than a star of the same mass.

A Tour of the Binary Star Bestiary

A low percentage of stars end their lives as neutron stars, and the percentage that end their lives as black holes is even lower: a few tenths of a percent. Black holes are as rare as black swans. To repeat: the distribution of the masses of stars as they form is highly skewed toward low-mass stars, and there are hundreds of low-mass red dwarfs for every Sun-like star. Red dwarfs die as fading embers called white dwarfs. Therefore, over 95% of all stars will end their lives as white dwarfs rather than as neutron stars or black holes.

Just over half of all stars are single, like our Sun, while a third are binary and 10% are in systems with three or more members.[35] Most binaries are in wide separation orbits with periods of years or decades or even centuries, so they don't interact and affect each other's evolution. A small fraction of binaries, less than 5% of the total, have orbital periods of between several hours and several weeks.

Any star has an imaginary boundary within which all material is gravitationally bound to it. For an isolated star, this boundary is a sphere. When binary stars are close together, these boundaries are stretched into teardrop shapes with the points touching. Mass can flow from one star to the other through the point where the teardrops connect. Typically, the higher-mass star will siphon gas off the lower-mass star. If they are really close together, the imaginary surfaces merge into a common envelope and mass can move easily between the stars.[36]

Most close binaries will contain two red dwarfs, since most stars are dwarfs. When these stars die, they collapse into white dwarfs, but low-mass stars live for so long that most of them have not died yet. High-mass stars have short lives, so if we locate binaries with a high-mass star and a low-mass star, it's likely that the more massive star will have died and left behind a neutron star or a black hole.

These are the types of stellar remnant binaries in order of increasing scarcity: double white dwarf, white dwarf and neutron star, white dwarf and black hole, double neutron star, neutron star and black hole, and double black hole. Let's call this last one the double black pearl; it's the rarest combination of all. We'll return to it later.

To tell all the stories of binary stars would require a book far longer than this one. Like the relationships between people, they're hugely varied. Couples can be large and small, and personalities can be hot and cold. There's give and take on both sides and one life profoundly affects the other. Sometimes one partner leaves the relationship, and one partner almost always dies before the other. With stars, a close relationship can even lead to life after death.

Consider two normal stars orbiting each other in the prime of life, fusing hydrogen into helium. The more massive star exhausts its hydrogen first and bloats into a red giant, spilling gas onto its companion. With both stars engulfed in gas, they spiral closer to each other. The more massive star dies, collapsing into a white dwarf. Eventually, the less massive star ages and swells, pouring gas onto its dead companion. The very strong gravity of the white dwarf compresses the gas sufficiently to ignite fusion, and it flares briefly back to life. This is called a nova, or "new star." The violent fusion ejects much of the gas, and the process can repeat episodically. Sometimes a nova takes a star from a faint glow in the telescope to naked-eye visibility.[37] If the mass transfer is large enough, the white dwarf can be pushed over the Chandrasekhar limit of 1.4 times the mass of the Sun. In this case, the dead star dies for a second time as a supernova, leaving behind a neutron star.[38]

Here's the life story of a binary that ends up with a black hole.[39] Two hot, massive stars live in a tight binary orbit. The more massive star exhausts the hydrogen in its core and it expands, shedding most of its envelope onto its companion, leaving behind a naked helium core. After a few hundred thousand years, it dies violently as a supernova that leaves behind a black hole. The less massive companion gains gas from the explosion, which accelerates its evolution. After 10,000 years it expands

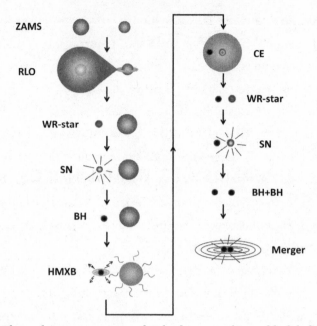

FIGURE 17. The evolutionary sequence that lead to a rare binary black hole system. At the top left the stars start on the Zero Age Main Sequence (ZAMS). The more massive star spills material onto its companion by Roche-lobe overflow (RLO). After some time in a Wolf–Rayet (WR) phase, the more massive star dies as a supernova (SN) then a black hole (BH). It emits X-rays as a high-mass X-ray binary (HMXB). The two stars then occupy a common envelope (CE). Then the second star dies similarly. The two black holes will eventually merge into one, more massive black hole. *Pablo Merchant/A&A, vol. 588, p. A50, 2016, reproduced with permission/copyright ESO*

as it reaches the end of its life, spilling gas onto the black hole and triggering intense X-ray emission. Then it too detonates as a supernova, and, depending on its mass, the final system includes either a neutron star and a black hole or a double black hole (Figure 17).

Black holes are a bizarre but inevitable consequence of the evolution of massive stars. If they are in binary systems, their interaction allows them to be detected. Every second, somewhere in the universe, a massive star dies violently. Every second, a patch of space-time is pinched off from view, and every second a black hole is born.

But what if there was another way to form a black hole? And what if the result was even more monstrous than anything previously imagined?

3.

SUPERMASSIVE BLACK HOLES

S A DEAD star the only possible type of black hole? The requirement for a black hole is density high enough to generate such strong gravity that no light can escape. In principle, this could occur in larger (and smaller) objects than a collapsed star. Nevertheless, it was a surprise when supermassive black holes were discovered, some so massive that they exceed the sum of all the stellar black holes in our galaxy. It was an even greater surprise to learn that one exists at the center of every galaxy.

The Only Radio Astronomer in the World

The summer of 1937 was hot and humid in Wheaton, Illinois. Twenty-six-year-old Grote Reber was out in the vacant lot next to his mother's home every day, cutting wood and shaping metal from seven in the morning until darkness fell. He was building a radio telescope. The dish was 10 meters in diameter, the biggest he could construct with available materials.[1] When completed, it was by far the largest radio telescope on earth—a precursor of modern versions that range up to 100 meters in diameter. For a decade, Reber was the only radio astronomer in the world (Figure 18).

But he wasn't the first. Karl Jansky was trained in physics and was

FIGURE 18. The first parabolic radio telescope in the world, built by amateur radio astronomer Grote Reber in 1937. The first radio telescope was Kark Jansky's dipole array. The 9-meter dish was built in Reber's backyard in Wheaton, Illinois. It was the prototype for all future dishes in the young field of radio astronomy. *Grote Reber*

just twenty-three when he was hired by Bell Labs in Holmdel, New Jersey. The company wanted to investigate the possibility of using radio waves 10–20 meters long for transatlantic telephone service. Jansky's job was to look into sources of static that might interfere with voice communications. In 1930, he built an antenna in a fallow potato field near the lab. The contraption looked like the wing frame of an early biplane, and it pivoted on a circular track on four rubber-tired wheels from a Model T Ford. By rotating the antenna, Jansky could tell the direction of incoming radio waves, while the radio signals were amplified and recorded by a pen on a moving paper chart in a nearby shack. Mostly, he detected static from nearby thunderstorms, but there was also a faint radio hiss. Within a year, Jansky had shown that the hiss was not of terrestrial origin; it kept the time of the stars, rising and falling not every 24 hours but every 23 hours and 56 minutes.[2] The radiation was strongest toward the center of the Milky Way, in the constellation of Sagittarius. Jansky's revelation caused a stir, and was reported by the *New York Times* on May 5, 1933.[3]

This was the birth of a new way of studying the universe.[4] For thousands of years of naked-eye astronomy, and hundreds of years since Galileo's first use of the telescope, all information from space had come in a slender range of optical wavelengths: a factor of only 2 from the red-dest red to the bluest blue the eye can see. Now humans had recorded signals from an entirely new region of the electromagnetic spectrum. Jansky proposed building a 30-meter radio dish so he could follow up his discovery. Bell Labs wasn't interested. They assigned Jansky to another project, and he did no more radio astronomy.

Jansky's work sparked Reber's curiosity about the source of cosmic radio waves. In the early 1930s, he applied to work with Jansky at Bell Labs, but during the Great Depression no one was hiring. So he taught himself how to build a telescope and a receiver. He enjoyed working alone. As he put it, "There have been no self-appointed pontiffs looking over my shoulder giving bad advice."[5] He settled into a rhythm. By day, he designed radio receivers at a factory in nearby Chicago. After supper, he grabbed four or five hours' sleep, then from midnight until sunrise his dish tracked around the sky while he sat in his basement recording the radio signal at one-minute intervals. Eventually, he improved his receiver and bought an automatic chart recorder so he wouldn't have to sit up all night. This enabled Reber to embark on the first survey of the radio sky.

Reber was isolated. He had no peers to exchange ideas with, and he was working at unexplored wavelengths. Imagine being the first sculptor in the world. Other people are painting and drawing but nobody else is creating artwork in three dimensions. With nobody else speaking your language, you'd be lonely. Engineers took little notice of Reber's work, which he published in a radio engineering journal as Jansky had before him. Astronomers, meanwhile, were uninterested or skeptical. After Reber confirmed Jansky's detection of radio waves from the Milky Way in 1940, he submitted a paper on what he called "cosmic static" to the *Astrophysical Journal*. The editor, Otto Struve, sent the paper to several referees. Engineers didn't understand the astronomical implications. Astronomers were confused by the radio jargon. Nobody was willing to recommend its publication. Struve decided to publish it anyway.[6] The only radio astronomer in the world continued his lonely work.

With meticulous care, Reber mapped the sky. He worked at successively shorter radio wavelengths, knowing they would yield more accurate locations for the radio emission. By working at several wavelengths he could diagnose the physical process causing the radiation. In 1944, he wrote a paper that included the first-ever map of the radio sky.[7] The paper also demonstrated that the emission process of cosmic radio waves was nonthermal, thus different from radiation that comes from an object at a fixed temperature. His maps showed that the radiation was concentrated in the Milky Way, with two other peaks of emission, in Cassiopeia and Cygnus. The first would prove to be a supernova remnant 11,000 light years away. The second, coincidentally not far from the prototypical black hole Cygnus X-1, would eventually prove to be a galaxy with immensely powerful radio emission half a billion light years away.

It would take some time for astronomers to understand the nature of this galaxy, dubbed Cygnus A, which is the strongest radio source in the sky (Figure 19). Its discovery established Reber as an iconoclast. As

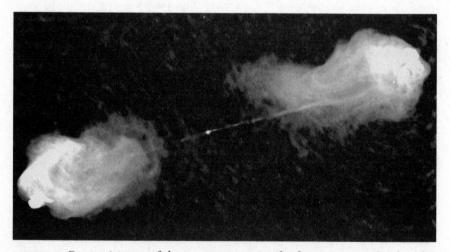

FIGURE 19. Cygnus A is one of the strongest sources of radio emission in the sky. This image was made with a radio interferometer called the Very Large Array. The bright radio spot at the center is now known to be a supermassive black hole in a galaxy 600 million light years away. Jets of high-energy plasma travel out of the core and create diffuse "lobes" of radio emission far beyond the galaxy. *R. Perley, C. Carilli, J. Dreher/NRAO*

he once advised a young student, "Pick a field about which very little is known and specialize in it. But don't accept all current theories as absolute fact. If everyone else is looking down, look up or in a different direction. You may be surprised at what you will find."[8]

Galaxies with Bright Nuclei

Science doesn't flow like a river. Rarely are scientists carried smoothly along to the sea of understanding. More often, they're explorers traversing difficult terrain—sometimes in daylight and with steady progress, sometimes in a fog with no compass. There are detours and dead ends. Different people working toward the same goal don't always communicate or even know of one another's existence. It's rare that someone is smart enough, or lucky enough, to find high ground and see the larger landscape.

At the turn of the twentieth century there was a fierce debate in astronomy over the nature of "nebulae," the fuzzy patches of light that had been cataloged by William Herschel (and others) over 100 years earlier. Since many had a spiral structure and they didn't sit close to the plane of the Milky Way, like most star-forming regions, astronomers started to take seriously the hypothesis that they were "island universes," separate systems of stars at enormous distances from our galaxy. If so, their spectra would look like the sum of the light of many stars, with the same absorption lines seen in the Sun and other stars. In 1908, Edward Fath at Lick Observatory looked at a spectrum of the nebula NGC 1068 and was surprised to see not just absorption but also six strong emission lines, features that could only be generated if the gas was heated by some source of extreme energy.[9] At the time, the result was so puzzling that it was ignored, and it would be two decades before Edwin Hubble proved that NGC 1068 was, in fact, a galaxy.[10]

In the early 1940s, Carl Seyfert was a postdoc at the Mount Wilson

Observatory in Southern California. With Edwin Hubble as his advisor, Seyfert used the 60-inch and 100-inch telescopes, the most powerful telescopes then available, for his research.[11] When Seyfert gathered his data, Los Angeles had a third the population and a tenth of the city lights that it does now. He also benefited from truly dark skies during the blackouts that were imposed after the attack at Pearl Harbor. He took spectra of the nuclei of bright galaxies and found half a dozen similar to NGC 1068, with bright emission lines indicative of an energetic process. He also noted that the emission lines were very broad. The width of an emission line indicates the range of motion of the gas. In a normal spiral galaxy, the maximum rotation speed is 200–300 kilometers per second; yet Seyfert was measuring Doppler widths of thousands of kilometers per second, suggesting that gas near the center of these galaxies was moving 10 or 20 times faster than anything previously measured. Material moving at this speed would fly away from the galaxy unless some great mass near the center was pinning it.

Seyfert had a puzzle to solve. What could cause rapid motion of gas in the center of a galaxy? At the time, nobody knew. Like Grote Reber's paper on "cosmic static" the following year, Seyfert's paper caused hardly a ripple in the astronomy world. It wasn't referenced for sixteen years after its publication.[12] The class of galaxies that was eventually named after Seyfert waited in limbo. Meanwhile, new insights came from technical advances in radio astronomy.

Radio Astronomy Comes of Age

The pure sciences seemed irrelevant in the 1940s, with the war on. However, many radio astronomers played a crucial role in the development of radar, which in turn was pivotal in the outcome of World War II. Battles were won by the first side to spot enemy airplanes, ships, or submarines. British and American engineers and scientists developed radar that could "see" for hundreds of miles, even at night. Radar helped to sink German U-boats, it allowed the British to spot incoming bombers, and it provided cover for the D-Day landings. It is often said that the atomic bomb ended the war, but radar won it.

Radar also led to astronomical discoveries. In 1942, Stanley Hey, of the Army Operational Research Group, was puzzling over severe interference in England's coastal radar defenses. He realized that the interference wasn't coming from the enemy, but from the Sun. Later in the war, he discovered the ionized trails of meteors while trying to track German V2 rockets. Neither discovery could be published until the war was over. His group also confirmed the existence and the strength of the puzzling radio source Cygnus A. After the war, Hey continued military work at the Royal Radar Establishment in southern England, while others who worked on wartime radar became the pioneers of radio astronomy. Martin Ryle started the Cavendish Laboratory at Cambridge University, and Bernard Lovell started Jodrell Bank as a field station of the University of Manchester.[13]

Australia became a radio astronomy power due to the technical expertise it contributed to the Allied war effort. One of the world's best wartime radar labs was located in Sydney, and after the war the lab stayed intact, with the staff switching to the study of cosmic radio "noise." Noteworthy among them was Ruby Payne-Scott, one of the best physicists Australia has ever produced and the world's first female radio astronomer. After contributing to the war effort, she was the first to study solar radio bursts, and she developed the mathematical formalism for the type of interferometry used at radio arrays all around the world. She battled sexism her entire career, and had to conceal her marriage since at that time married women were not allowed to hold full-time jobs as civil servants.[14]

Meanwhile, European astronomy was kick-started at war's end by the redeployment of 7.5-meter antennas from German radar facilities to national observatories in Britain, the Netherlands, France, Sweden, and Czechoslovakia. This is a gratifying story of swords being turned into scientific ploughshares.

In 1946, Stanley Hey and his colleagues used their modified anti-aircraft radar antenna to show that Cygnus A varied in strength from minute to minute. Since light can only travel a certain distance in such a small amount of time, any variation timescale sets a size scale for a source of radiation. Fast variations can only be seen in very small sources of radiation. In this case, the object was determined to be very small, the size of a star. Martin Ryle suggested that Cygnus A might be a new kind of star that was bright at radio wavelengths but invisible at optical wavelengths: a "radio star." This confused everyone.[15] Stars like the Sun emit feeble amounts of radio waves, so how could a star be such a bright radio source? As the radio astronomer J. G. Davies commented, "There appeared to be an optical universe and a radio universe which were utterly different, yet which coexisted. So there was obviously a need to tie them together somehow."[16]

The obstacle to progress in radio astronomy was angular resolution: the smallest angle any telescope can distinguish. Better angular resolution corresponds to a smaller angle. If light sources are closer together than the angular resolution of a telescope, they blur together. Angular resolution also affects the depth of vision; when sources blur into one another it's impossible to tell which is nearer and which is farther. Imagine being myopic in a large room full of people. You may make out the nearest few faces, but all the rest are hopelessly blurred. It would be difficult even to count the people in the room. Put on glasses, and everything snaps into focus.

Sharper images require either shorter waves or a bigger telescope.[17] Angular resolution is proportional to the wavelength of observation and inversely proportional to the telescope's size. Radio waves are millions of times longer than light waves, so radio astronomers start out at a serious disadvantage in comparison to their optical counterparts. They compensate for that in part by building big telescopes. Grote Reber's dish was

9.4 meters in diameter, larger than any optical telescope at the time. Its sharpest images were 15 degrees across. That's the width of a fist held at arm's length. There are many optical objects in a patch of sky that big, so it was impossible for Reber to deduce the source of the radio waves. Shifting to higher frequencies, which means shorter wavelengths—200 cm rather than 2 meters—a factor of 10 could be gained. To put this in perspective, visible light waves are 3 million times smaller than the 2-meter-long waves Reber observed. An optical telescope the same diameter as Reber's would make images 3 million times sharper. To produce an image as sharp as a 1-meter optical telescope would require a radio telescope the size of the United States!

The invention of interferometry solved the problem. In an interferometer, incoming waves from two (or more) radio telescopes are combined with the phase information of the waves preserved—meaning, the exact timing of arrival of the peaks and troughs. Thus, angular resolution is given by the separation of the telescopes rather than by size, so that two 10-meter dishes separated by one kilometer have 100 times better angular resolution than either dish used separately.[18] This technique is also called aperture synthesis, because it "synthesizes" a telescope with the resolving power of a very much larger telescope. In 1950, Graeme Smith at the Cambridge's Cavendish Laboratory used two repurposed German antennas to measure the position of the bright radio source Cygnus A with an accuracy of one arc minute, or one thirtieth of the Moon's diameter—a hugely important breakthrough (Figure 20).

The precise measurement of the position of Cygnus A got the attention of Caltech astronomer Walter Baade. Within two weeks of receiving the data from Smith, Baade was in the observing cage of the 200-inch telescope on Mount Palomar, the most powerful optical telescope in the world. The German-born astronomer hadn't been allowed to enlist during World War II, so like Carl Seyfert he used the wartime blackouts of Los Angeles to take photographs of unprecedented depth of

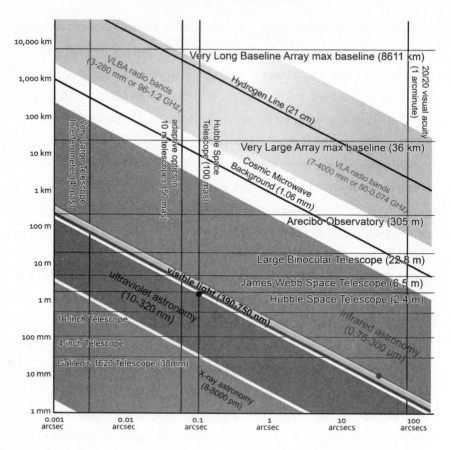

FIGURE 20. A logarithmic plot of telescope size versus angular resolution, or the ability to resolve small features in an astronomical object. For reference, the angular size of the Moon is 1,800 arcsecs. The dot at 10 mm and 20 arcsecs is the unaided eye, and the dot at an aperture of 2.4 meters and a resolution of 0.1 arcsec represents the Hubble Space Telescope. The wavelengths of electromagnetic radiation are shown as diagonal lines, increasing from X-rays at the lower left to radio waves at the upper right. The largest single telescopes are about 10 meters at optical wavelengths and about 300 meters at radio wavelengths. Interferometers simulate a very large telescope, so give a very high angular resolution. *Chris Impey*

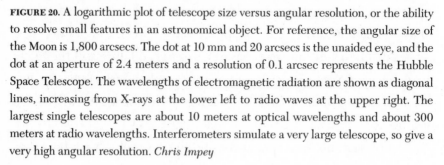

the night sky. A vignette from a later interview paints a vivid picture: "When he told what he had seen and discovered in careful scanning of thousands of plates, the incredible grandeur of the cosmic realm, the galactic and extragalactic world, began to unfold behind numbers, pictures, and astronomical gossip. This man in his dark blue tie and gray suit, wearing brown shoes of enormous size, was absolutely fascinated by his research. Gesticulating, incessantly smoking, with carefully parted thin white hair, white somewhat bushy eyebrows, protruding hawk nose, Baade saw the mysteries of the universe as the greatest of all detective stories in which he was one of the principal sleuths."[19]

When Baade pointed the 200-inch telescope, nicknamed the Big Eye, at Cygnus A, the resulting photographic plates thrilled him: "I knew something was unusual the moment I examined the negatives. There were galaxies all over the plate, more than 200 of them, and the brightest was at the center. It showed signs of tidal distortion, gravitational pull between the two nuclei—I had never seen anything like it before. It was so much on my mind that while I was driving home for supper, I had to stop the car and think."[20] With the cooperation of radio and optical astronomers, one important question had been answered: Cygnus A was a distant, distorted galaxy. Its spectral lines were redshifted by 5.6%, indicating that it was receding at 35 million mph. In terms of an expanding universe model, where redshift indicates distance, that meant it was 750 million light years away. The radio waves we see now were emitted when life forms on Earth were no larger than the head of a pin.

Baade had been thinking of the power that might result from cosmic "train wrecks," and proposed that super-bright Cygnus A comprised a pair of colliding galaxies. Rudolph Minkowski, one of Baade's colleagues at Caltech, questioned his theory, and Baade tried to bet him $1,000 that his hypothesis would prove correct. (Black hole theorists evidently aren't the only scientists prone to gambling.) At the time, that was a month's salary. Minkowski wouldn't bite, so Baade lowered the bet to a bottle of whisky. Minkowski conceded the bet after he took a spectrum that showed emission lines from very hot gas near the center of Cygnus A. When two galaxies collide, the gas they contain heats up. (Baade later griped that Minkowski paid off the bet with a flask of whisky rather than a bottle, and then drained the flask on a subsequent visit to Baade's office.) Nonetheless, a number of theorists later decided, through careful calculations, that a collision couldn't explain the radio brightness. A central question remained unanswered: how *could* Cygnus A emit 10 million times more radio waves than the Milky Way?

A Dutch Astronomer Discovers Quasars

In 1950, a new mechanism was proposed for cosmic radio waves.[21] When electrons move at close to the speed of light in a magnetic field, they travel in a spiral pattern and emit intense radiation over a broad range of wavelengths. This is called synchrotron radiation. Synchrotron radiation was created in lab accelerators in the 1940s, but it was a surprise to learn that this process could occur when particles are accelerated in regions of space hundreds or thousands of light years across. In 1958, at an international astrophysics conference in Paris, scientists presented papers arguing that synchrotron radiation could explain solar flares, afterglow from a 1054 AD supernova in the Crab Nebula, the peculiar elliptical galaxy M87—and possibly Cygnus A.

When Cambridge radio astronomers published their third catalog in 1959, optical astronomers targeted the most compact radio sources, which offered the best chance of finding an optical counterpart.[22] As before, Caltech astronomers were at the center of the action. Tom Matthews and Allan Sandage observed the 48th object in the catalog, 3C 48, and found a faint blue object at the radio position, surrounded by wisps of nebulosity. The light varied rapidly; the object couldn't be much bigger than a star. Most puzzling of all was the spectrum: it had strong and broad emission lines that couldn't be identified with any known element. Matthews showed it to Jesse Greenstein down the hall, but Greenstein—an expert on stars—had never seen a stellar spectrum like it. Unable to explain the data, Greenstein put the spectrum in a drawer and forgot about it.

It fell to Maarten Schmidt to take the next step. The young Dutch astronomer came to Caltech to work on star formation in galaxies, but he was intrigued by the puzzle of radio sources. In 1963, radio astronomers in Australia took advantage of the Moon eclipsing the 273rd object in the 3C catalog to measure a very accurate position for the source. Schmidt went to the 200-inch telescope and saw a blue starlike object with a linear feature off to one side. A spectrum of the object showed mysterious emission lines like those seen for 3C 48. When Schmidt attempted to match the pattern of lines with a well-known element, he realized the features were hydrogen lines shifted 16% to the red. Taking redshift as due to cosmic expansion, 3C 273 was moving away from us at the incredible speed of 100 million mph. Four classic papers describing Schmidt's discovery were published in the journal *Nature*.[23]

Fifty years later, he recalled the moment of discovery: "Interpreting it as a cosmological redshift, which I soon did because it was so bright in the sky—the luminosity turned out to be very high. And that was remarkable, because it was immensely brighter than normal galaxies, even the biggest galaxies. So here you have something that is out in the universe, more luminous than an entire galaxy, and it looks like a star. It was an astounding experience" (Figure 21).[24]

With the help of Schmidt's insight, Greenstein dug up the previous spectrum of 3C 48 and quickly identified lines from normal elements redshifted more than twice as much, by 37%.[25] 3C 273 was 2 billion light years away and 3C 48 was 4.5 billion light years away; light from the latter was emitted just as the Earth was forming and has been traveling through space ever since. Both objects were emitting 100 times the light of an entire galaxy, yet were smaller than the Solar System. Maarten Schmidt coined a new term for this extraordinary class of objects: quasi-stellar radio source, or quasar.

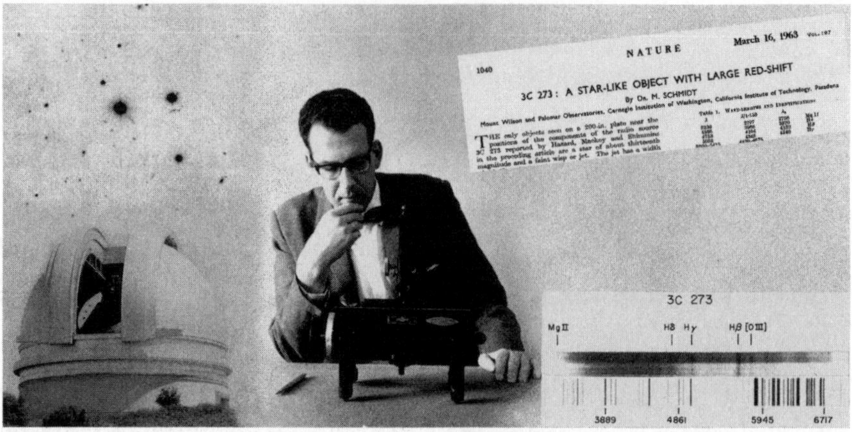

FIGURE 21. Montage created for a symposium commemorating the fiftieth anniversary of Maarten Schmidt's discovery of quasars. Left: The Palomar 200-inch telescope and the image of the quasar 3C 273. Middle: Maarten Schmidt inspecting the spectrum that was taken with the telescope. Right: The spectrum of 3C 273, with emission lines marked above and wavelengths marked below, along with the title page of the paper that announced the discovery. *G. Djorgovski/California Institute of Technology*

Since the pivotal work was done in Los Angeles (Pasadena having been subsumed by the sprawl of Los Angeles long ago), let's use the greater Los Angeles metropolis as an analogy. Imagine you're hovering high above Los Angeles at night in a helicopter. In the city of roughly 10 million there are approximately ten house, street, and car lights for every person, for a total of about 100 million lights (I'm rounding these numbers to the nearest order of magnitude for simplicity's sake). If Los Angeles were a galaxy, each light would represent about 1,000 stars. Now imagine that in downtown Los Angeles there's a single-point source of light that shines several hundred times brighter than the entire city, yet it's only a few inches in size, no bigger than any of the individual lights. If we could rise high above the Earth, so that the city was thousands of miles below us, the intense central light source would remain visible long after the individual lights had faded from view. From a great distance across the universe, a galaxy can be too small and faint to see, but its bright core shines brilliantly. That's a quasar.

Astronomers Harvest Distant Points of Light

The most striking property of quasars is their high redshift, indicating large distances and high luminosities. The expansion of the universe stretches the wavelengths of photons that travel through it, and that effect is called the *cosmological* redshift.[26] Redshift, written z, is defined as $1+z = R_o/R_e$, where R_o is the size of the universe (or the distance between any two points in space) when the light from an object is observed, and R_e was the size of the universe (or the distance between any two points in space because all space expands at the same rate) when that same light was emitted. It's exactly the same relationship for radiation, $1+z = \lambda_o/\lambda_e$, where λ_o is the stretched or reddened wavelength of a photon we observe now with a telescope, and λ_e is the wavelength of that photon when it was originally emitted.

The farther away a galaxy is, the faster it's receding; in fact, every galaxy is moving away from every other galaxy.[27] That observation, made by Edwin Hubble in 1929, led to the idea of an expanding universe. When redshift is small, it's roughly equal to the recession velocity as a fraction of the speed of light.[28] Before quasars were discovered, the most distant object known was a galaxy in the Hydra cluster, at a redshift of $z = 0.2$. Within two years Maarten Schmidt had pushed the redshift record to $z = 2.0$ with 3C 9,[29] a quasar receding at 80% of the speed of light. The light we see now was emitted when the universe was a quarter of its current age (Figure 22). Since distant light is old light, astronomers use faraway objects as "time machines." Quasars are probes of the distant and ancient universe.

Finding quasars was hard work in those early days. Radio astronomers had to grind to get accurate positions. A typical day at the telescope would be made up of two twelve-hour shifts, since radio waves can be equally well detected by day and by night. The maze of electrical connections in the control room had to be checked and double-checked.

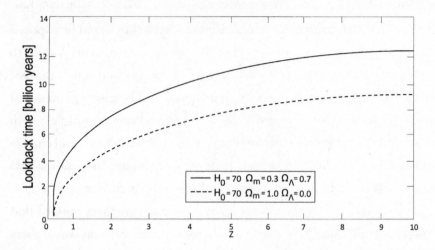

FIGURE 22. Distant light is ancient light. Redshift is caused by the expansion of the universe, and Hubble showed that redshift increases with distance from the Milky Way. The graph shows the relationship between redshift, or the fractional wavelength change due to light travel in an expanding universe, and lookback time, or how long ago the light was emitted. The universe is 13.8 billion years old. The light of quasars has been traveling for a substantial fraction of the age of the universe. The solid core shows the currently accepted chronological model. *Chris Impey*

Plugs would connect the signals from different telescopes or parts of an array into a correlator, and only an expert could keep track of this spaghetti junction. These were the very early days of computing, so signals were recorded in analog fashion on magnetic tape. It was a full-time job to monitor the tape decks and spool up new tapes, making sure the tapes never ran out. The signals were then fed into a mainframe computer on punch cards, to monitor the changing strength of the radio signal as the source moved across the sky and combine it with accurate clock readings to calculate a position. Many days and many sweeps of the sky were needed to get a single accurate position.

The life of an optical astronomer was slightly easier and more glamorous. They rode the prime focus cage of a large telescope, suspended above the primary mirror like a fly caught in a metal web. The open slit of the dome looked out onto glittering star fields. They would take unexposed photographic plates into the cage, sealed in a light-tight box, and carefully transfer them to the camera where they could be exposed to the night sky. Then they would use buttons on a small paddle to make fine adjustments to the rate the telescope tracked across the sky, to ensure the images were as sharp as possible. There was glamor, but tedium too. In winter, there were cold, twelve-hour-long nights, with not much to do but push guiding buttons every few seconds and change plates every few hours. Optical astronomers might have to spend a whole night at a telescope to measure the redshift of a single object.

With just a few dozen quasars cataloged, astronomers noticed that quasars were bluer (that is, hotter) than any star. Several researchers realized that there were other, equally blue quasi-stellar objects that weren't associated with any radio source. Spectra revealed that many of these blue objects had high redshifts; they were quasars too. Excited by this discovery, astronomers did photographic surveys over large swathes of the sky to "harvest" objects with the bluest colors. The method was highly effective: quasars selected in this way outnumbered quasars with strong radio emission by a factor of 10.

Occasionally, the competitiveness of the hunt for quasars spilled into acrimony. In 1965, Allan Sandage at the Carnegie Institution wrote a paper about the new class of radio-quiet quasars, and such was his reputation that the editor of the *Astrophysical Journal* published the paper without peer review. Fritz Zwicky, at Caltech, was incensed, as he'd previously discovered compact galaxies with quasar properties. A few years later, he wrote in the vituperative foreword to his book on the properties of peculiar galaxies: "In spite of all these facts being known to him in 1964, Sandage attempted one of the most astounding feats of plagiarism by announcing the existence of a major new component of the Universe: the quasi-stellar galaxies. Sandage's earthshaking discovery consisted of nothing more than renaming compact galaxies, calling them 'interlopers' and quasi-stellar galaxies, thus playing the interloper himself."[30] So much for genteel academia.

Competition between Carnegie and Caltech led to many ambitious projects that are driving optical astronomy in the twenty-first century. First, Caltech built the twin 10-meter Keck telescopes in Hawaii, while Carnegie built the twin 6.5-meter Magellan telescopes in Chile. Carnegie is now lead partner in the 22.5-meter Giant Magellan Telescope, and Caltech is lead partner in the planned Thirty Meter Telescope.[31] Both are billion-dollar projects with international partners. As astronomers seek to harvest ever more distant points of light, their "toys" are getting more complicated and much more expensive.

The University of Arizona is making the mirrors for the Giant Magellan Telescope. Every year or so, I visit the facility under the football stadium where twenty tons of pure glass in small chunks are put into a tub 30 feet across, heated to 1170° C, and spun into a parabolic shape. When the huge mirror oven is spinning, with flashing lights and waves of heat pouring out, it's like a sinister fairground ride, and the engineers nearby, wearing white coats and safety goggles, look like mad scientists. Three months later, when the mirror has fully cooled, it's polished to near-perfection. It amazes me to think that if the finished mirror was expanded to the size of the continental United States, the largest undulations or blemishes would be under an inch high. The Giant Magellan Telescope uses seven large mirrors, with six of them arrayed like flower petals around the central element. Meanwhile, the Thirty Meter Telescope will be constructed with 492 hexagonal mirrors, each five feet across. Both projects are racing to be the new largest telescope in the world. Each will spend a significant fraction of its time studying quasars.

Hypothesizing Massive Black Holes

Even before quasars were discovered, there were reasons to believe something very unusual was going on in the center of some galaxies. In 1959, a calculation showed that the broad emission lines in Seyfert galaxies could be explained by gravity from a compact object a billion times more massive than the Sun. English theorist Geoff Burbidge succinctly stated the challenge of radio galaxies: the energy they contain in magnetic fields and relativistic particles requires the complete transformation of up to 100 million solar masses into energy.[32] Relativistic particles are particles moving at close to the speed of light. Armenian theorist Victor Ambartsumian proposed "a radical change in the conception of the nuclei of galaxies," saying, "we must reject the idea that the nuclei of galaxies are composed of stars alone."[33]

Hypotheses swirled. Perhaps the energy resulted from explosions in a dense star cluster, as one supernova set off others in a chain reaction. Perhaps a star cluster could evolve to very high densities by collisions ejecting large amounts of gas. Perhaps the energy came from a single, supermassive star. Then, within a year of Schmidt's dramatic discovery, a pair of theorists suggested that the source of quasar power was accretion onto a supermassive black hole.[34] They recognized that stellar fusion was too inefficient to generate quasar power. A gravitational engine is needed. With mass spiraling into the innermost stable orbit of a massive black hole, mass could be converted into particle and radiation energy with an efficiency close to 10%. Even at this efficiency, the most luminous quasars would require black holes a billion times the mass of the Sun to power them.

The astrophysics community didn't immediately swoon for supermassive black holes. Remember that 1964 was the year in which the term "black hole" was coined and Cygnus X-1 was first detected. The idea of stellar-mass black holes was still a novelty, yet here were theorists suggesting black holes a billion times more massive! It sounded like wild speculation. Can you imagine a factor of a billion? It's the difference between one grain of sand and a sandbox full. It's the difference between having enough money to buy a burger, and being the richest person in the world. It's the difference between the mass of the people in your immediate family and the mass of Mount Everest. Even seasoned astrophysicists were taken aback by the thought of black holes as massive as small galaxies.

The extreme energy requirements of quasars hinge on the fact that they are at large distances from the Earth and thus must have very high luminosities to be as bright as they are. Luminosity is an intrinsic brightness, or how many photons the source emits every second. If quasars weren't at the distances indicated by their redshifts, then those energy requirements would be eased. The logic goes like this. A 100-watt lightbulb at a distance of 100 meters would appear faint. But if the lightbulb was actually 100 kilometers away, it would have to be a million times more luminous to appear to you to have the same brightness, making it a 100-megawatt lightbulb. Quasars are faint but so far away—billions of light years—that they must be fantastically luminous.

This issue led a small but vocal group of astronomers—including some highly respected names—to question the cosmological nature of quasar redshifts.[35] The cosmological redshift in an expanding universe model translates into distance. These astronomers pointed to places where quasars were seen near galaxies at much lower redshift; there were more of these than should happen by chance. They noted an excess of specific redshifts that had no explanation in the cosmological interpretation. These statistical claims were not compelling to most astronomers, but arguments based in the physics of energy density were more troubling. Physicists argued that quasars would "choke" on their own radiation and be quenched before they could ever shine brightly. Quasars with very rapid radio variability were so compact that when relativistic electrons emitted radio photons they would hit those photons and boost them to optical, then X-ray, then gamma ray frequencies. The result would destroy the radio source and turn it into a gamma ray source. In the mid-1960s there were many heated debates on this topic at conferences, with no consensus. It took new and better radio observations to make progress.

Mapping Radio Jets and Lobes

Radio astronomers could be forgiven for feeling a little peeved. They'd provided the first evidence of extreme energies in galaxy nuclei and the accurate positions that allowed quasars to be discovered. But quasars can't be understood without measuring a redshift, which requires an optical spectrum, and most quasars turn out to have feeble radio emission. It seemed like all the action was in optical astronomy.

But radio astronomers had another trick up their sleeve. In the discovery phase of quasars they had used dishes separated by hundreds of meters to get positional errors of about one arc minute. However, by increasing the separation in an interferometer to a kilometer and using as short a wavelength as possible, they reached accuracy of an arc second, close to the accuracy of an optical position. They could map the radio sky as accurately as optical astronomers. Seen in this detail, radio sources were amazingly diverse. There were radio galaxies, where the optical counterpart was obviously a galaxy, and quasars, where the optical counterpart was starlike. The most commonly seen type of radio source has huge lobes of radio emission straddling an elliptical galaxy with radio emission in its core, and in some cases the lobes extend several million light years into intergalactic space.[36] The galaxy often has a peculiar or disturbed morphology. It looks like beams of high-energy particles are being ejected from the center of the galaxy and powering the radio glow in the double lobes. Cygnus A is a beautiful example.[37]

We've encountered galaxies with intriguing and unusual properties. Some have strong radio emission; others have strong X-ray emission. Some have intense optical emission and rapidly moving gas near their centers. None of this behavior is characteristic of a galaxy that's simply a large collection of stars. Astronomers use the term "active galaxies" to refer collectively to galaxies where the nuclear regions are particularly energetic.

Because I'm an optical astronomer, I generally prefer data I can see—but to understand active galaxies I've used the Very Large Array in New Mexico, working in the same control room where Jodie Foster heard a message from aliens in the movie *Contact*. The VLA is a set of 27 dishes each 25 meters in diameter that can be arranged in a Y-shaped configuration spanning 25 miles. The dishes move on railroad tracks to make the separations between dishes larger or smaller. It took a while to get familiar with the jargon of radio astronomy. While the local radio astronomers readily helped me reduce my data, I noticed that they liked to retain an aura of mystique about their subject. I was at best an honorary member of the tribe.

Radio astronomers focused particular attention on the sources that were unresolved by the existing interferometers, where variability indicated the source wasn't much bigger than our Solar System. In the 1960s, they set about making a radio telescope as big as the Earth. They had to figure out a different way to combine the signals from different telescopes, since cable and microwave links wouldn't work on transcontinental scales. Their method was to record the signal from each telescope on magnetic tape, with the time recorded by an atomic clock, and then later bring the tapes together to create interference fringes and eventually a map. Data reduction was tedious and depended on advances in atomic clocks, computers, and magnetic tape recorders. In 1967, groups in the U.S. and Canada observed several sources over a distance of 200 km. Within a year, they added increasingly remote antennas in Puerto Rico, Sweden, and Australia. The baselines went up to 10,000 km, or 80% of the Earth's diameter. Angular resolution improved by a factor of 1,000 to a thousandth of an arc second—the angular resolution of a dime on top of the Eiffel Tower as seen from New York City (Figure 23). Radio astronomers could now produce images that were much sharper than those of optical astronomers.

This new technique was called Very Long Baseline Interferometry, or VLBI. In 1970, radio astronomers studying quasars using VLBI

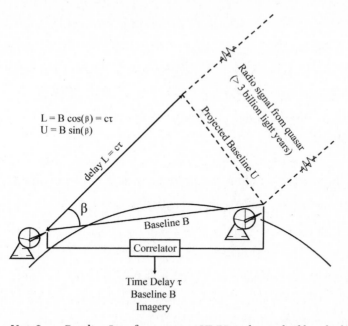

$L = B \cos(\beta) = c\tau$
$U = B \sin(\beta)$

delay $L = c\tau$

Radio signal from quasar (> 3 billion light years)

Projected Baseline U

β

Baseline B

Correlator

Time Delay τ
Baseline B
Imagery

FIGURE 23. Very Long Baseline Interferometry, or VLBI, is the method by which signals from widely separated radio telescopes are combined to simulate the very high angular resolution of a telescope the size of the maximum separation between the telescopes. Light arrives from a distant quasar at slightly different times at two telescopes with a time delay that's determined by simple geometry. The separate radio signals are combined by a piece of electronics called a correlator. *Chris Impey*

noticed that the most compact radio sources had one-sided jets, and there were often "blobs" or hot spots within the jets. In data gathered over the course of a year, they could see the blobs moving away from the nucleus. Astronomers are used to the huge timescales of the extragalactic universe, where it takes a galaxy hundreds of millions of years to rotate once, so it was gratifying to see changes from year to year.[38] But when they converted the apparent transverse motion of the blobs into a speed, they got a shock: the separation speeds were 5 to 10 times the speed of light. Was this a violation of relativity? No, it was an optical illusion. Because the jet in compact radio sources is pointing nearly at us and the blobs are moving at close to the speed of light, they can appear to have very rapid transverse motion. It's like someone on Earth using a powerful light beam to sweep a spot of light across the Moon's surface. If the beam pivots quickly enough, someone on the Moon could see the spot appear to move faster than light even though the photons in the light beam are traveling at light speed and no quicker. This phenomenon, called superluminal motion, has been seen in dozens of compact radio sources.

Exquisite mapping of radio sources showed that radio astronomers can make images just as beautiful as images in optical astronomy (Figure 24).[39] The data support the hypothesis of supermassive black holes. Strong radio emission implies that a particle accelerator is operating and the compactness means the radiation comes from a tiny region of space. Only a gravity engine such as a black hole can do this. Also, since galaxies have angular momentum and a compact object at the center of a galaxy should be spinning, gas will escape along the poles of the spin axis. A

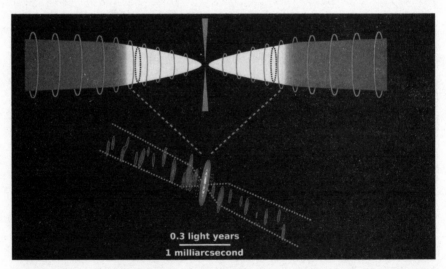

FIGURE 24. Data on the active galaxy NGC 1052 from a radio interferometer working at millimeter wavelengths (below), with a sketch of the horizontal jets and a cross-section through the vertical accretion disk (above). A magnetic field helps align and power the jets, and the data allows the field strength to be measured near the event horizon of the central black hole. *Ann-Kathrin Baczko/A&A, vol. 593, p. A27, 2016, reproduced with permission/copyright ESO*

black hole can be a particle accelerator far more powerful than the best man-made machines. Gravity powers twin jets of magnetized plasma that shoot out from close to the black hole at nearly the speed of light, extending far beyond the edge of the galaxy to illuminate the radio night.

The Zoo of Active Galaxies

In the parable of the elephant and the blind men, each man touches an elephant to see what it's like. One feels a leg and says the elephant is like a pillar. Another feels the tail and says it's like rope. A third touches an ear and says it's like a palm frond, and a fourth grasps a tusk and says it's like a pipe (Figure 25). The parable illustrates the dangers of inference

FIGURE 25. In this Japanese print from the nineteenth century, blind men examine an elephant, and each comes to a different conclusion regarding the nature of the animal. The image acts as a metaphor in science for the dangers of incomplete information, and in astronomy for the difficulty of combining information from different parts of the electromagnetic spectrum. *Itcho Hanabusa*

using incomplete information. With animals in mind, let's look at the "zoo" of active galaxies.

Active galaxies are defined by a negative: they show energetic behavior that can't be explained by stars or stellar processes. The subject started with the spiral galaxies that Seyfert discovered in 1943. Their bright blue nuclei and broad emission lines implied gas moving too fast to be explained by the normal rotation pattern of a galaxy.[40] In hindsight, it's clear that Seyfert galaxies are the "missing link" between normal galaxies and quasars, because they have nonthermal emission but are closer and less luminous than quasars. However, since Seyferts had been forgotten for several decades, quasars seemed unprecedented when they were discovered. Using the Hubble Space Telescope, astronomers made deep images and showed that the "fuzz" around quasars was actually the light of a distant galaxy. In a manner reminiscent of the analogy of Los Angeles at night, this demonstrated that the quasar light source really did live in a city of stars.[41]

There was a similar attempt to classify the different species of radio sources. Radio galaxies with low luminosity had cores and twin jets that typically ended within the galaxy in irregular lobes of emission. Radio galaxies with high luminosity had cores and one-sided jets extending to lobes far beyond the host galaxy. The radio sources with the strongest cores were quasars, with rapidly variable radio and optical brightness and extremely high energy densities. The most extreme species of all were called "blazars." As suggested by the moniker, they have dramatic brightness variations, sometimes changing from minute to minute. Their properties are consistent with situations where we're looking down the throat of a relativistic jet toward the central engine: the supermassive black hole.[42]

Several decades ago, I was in Russia to hunt blazars, but got more than I bargained for. Sometimes, the trip seemed straight out of a spy novel. My misgivings started when a pair of heavily built men got into the backseat of the car on either side of me, toting guns. The life of an observational astronomer isn't usually so eventful. We were heading across the border to an ice cream factory in Georgia to barter for dry ice to cool the instrument we'd brought from the U.S.

We'd come to the Russian 6-meter telescope, the world's largest, to study the rarest beast in the extragalactic zoo. The light from a blazar can vary by 100 times the luminosity of an entire galaxy in less than an hour. Our instrument was a photometer that could measure the brightness of a distant source of radiation in less than a second. We were hoping for an unobstructed view into the maelstrom near a supermassive black hole. My partner in crime was Santiago Tapia, a Chilean astronomer whom I'd met in Arizona. Our hosts were the staff at the observatory, senior PhD scientists paid less than $100 a month, who struggled to buy clothes and feed their families. As a young postdoc working in America, I was rich by comparison.

This was Russia in the waning days of the Soviet Union. Signs of stress and decay were everywhere. In Leningrad we'd seen empty shelves in the markets and long lines for the few restaurants. On the three-day journey to the Caucasus, where the telescope was located, soldiers stormed the train brandishing Kalashnikovs to look for bands of thieves. Naively, I went for a hike high above a river valley the day after we arrived. That evening, as we ate watery borscht and dense bread with our hosts, they said I should be careful because Georgian gunrunners used the valleys, and they could be unpredictable.

We gathered data the hard way. Santiago and I took turns riding the prime focus cage, a metal cylinder at the top of the telescope where the light came to a focus after bouncing off the primary mirror. The cage also housed the photometer we'd brought with us from the United States. The cage had no padding, and despite my layers of winter clothes I was chilled to the bone by the end of a long February night. But there were moments of exhilaration. One clear night our target flickered into action and the instrument's photon counter saw its light surging and falling. I imagined a star being torn asunder as it hit the accretion disk and became fuel for the beast. At the end of the night we sat with our Russian hosts and ate "poor man's caviar," made of finely diced pickled vegetables. We finished a bottle of rough-edged vodka and told stories until a swollen red Sun rose over the Caucasus.

The "elephant problem" for active galaxies is caused by selective vision. If you look with radio methods, you'll see a core and jets and lobes, but the majority of active galaxies are radio-quiet. If you look with optical methods, you'll see broad emission lines and a bright core with a faint surrounding host galaxy, but miss the jet phenomenon. These two slivers of the electromagnetic spectrum can't tell the whole story. We need other ways of looking.

As we've seen, X-ray astronomy led to the discovery of the pro- totypical black hole Cygnus X-1 in 1964. Six years later, a rocket detected X-rays from two nearby active galaxies, Centaurus A and M87, and a quasar, 3C 273.[43] In the 1970s, the orbiting Einstein Observatory had the sensitivity to detect large numbers of quasars. Their X-rays were variable, showing that they came from near the central engine. The ultraviolet and X-ray emission of many quasars looked like thermal radiation from gas at a temperature of 100,000 Kelvin. Excitingly, it matched the models of an accretion disk around a supermassive black hole.[44]

Each time astronomers opened up a new wavelength window, active galaxies were detected. The Infrared Astronomical Satellite, launched in 1977, found that quasars were strong infrared emitters. The hunch was that short-wavelength radiation created near the nucleus was repro- cessed by dust grains farther out into longer-wavelength infrared radi- ation.[45] During the 1990s, NASA's Compton Gamma Ray Observatory added a high-energy window onto active galaxies. The twin jets that emerge from the poles of a black hole can crank out a large amount of gamma rays. Certain active galaxies have been observed over an incredible factor of 100 million trillion (10^{20}) in wavelength, from meter-length to smaller than an atomic nucleus. In 2018, a spectacular new window onto active galaxies was opened up when a neutrino was detected from a blazar 4 billion light years away. Until then, neutrinos had only been detected from the Sun and a relatively nearby supernova. The neutrino was created near the blazar's central supermassive black hole and detected 4 billion years later by an array buried in the Antarctic ice.[46]

The elephant problem can be exacerbated by wavelength chauvinism. Astronomers specialize not only in the objects of their attention but also in their methods of observation. Optical astronomers—still the majority of all professionals—are following the classic trajectory of the subject, from naked eye to photographs to CCDs. Radio astronomers often come from an engineering background, and infrared and X-ray astronomers often come from a physics background. Beyond the technical distinctions, there's a "tribal" aspect to astronomers working at different wavelengths. Sometimes they don't talk to one another when they should.

A Matter of Perspective

Astronomers attempt to unify the various species in the "zoo" of active galaxies by supposing that their appearance depends on orientation. Spiral galaxies are flattened and accretion disks are thin, so it would be expected that the properties of active galaxies would depend on their orientation in space. To use a simple analogy, a sphere always has a circular shape regardless of orientation, while a thin disk might look like a circle, an ellipse, or even a line depending on how it is oriented.

Radio astronomers realized that the difference in radio brightness among quasars might not be due to intrinsic differences in luminosity. If jets accelerating particles to nearly the speed of light were oriented close to the line of sight, their emission would be dramatically boosted.

A direct view down the polar axis of the supermassive black hole would show a strong radio core, a one-sided jet, and perhaps a weak halo of extended radio emission. These are the rapidly variable blazars—which make up a small fraction of the overall population, since that orientation is very particular.[47] A side-on view of that same source shows a weak core, twin jets, and extended lobes on either side.[48]

I worked on blazars for my PhD thesis and for a decade after. They had the attraction of a sports car to a young man—fast and finicky and just as liable to leave you stranded on the road as to give you a thrilling ride. Blazars are unpredictable because their emission depends on changeable astrophysics close to a supermassive black hole. Sometimes I went to the telescope and almost all of my favorite targets were biding their time, and some were too faint to observe at all. But when I got lucky, blazars gave me a *Guinness Book of Records* payoff. At different times, I bagged active galaxies with the highest luminosity, the most rapid variations, the most compact emission, and the highest polarization. Polarization occurs when the vibrations associated with electromagnetic radiation are in a single plane; the polarization of light gives information on the geometry of the source of the light.

Good science, however, requires an analytic approach and systematic observations. So my research advanced by the totality of the data rather than by the most exciting moments. I learned that blazars represent a very privileged view onto the central engine. Hot gas moving at 99% of the speed of light means that blazars are hundreds of times brighter than an active galaxy when one is not looking along the jet. Accelerating gas to this speed was a theoretical challenge, but observers like me enjoy taxing theorists. It was eventually possible to identify the much more abundant active galaxies whose behavior was less exciting, representing the view when we're not looking down the jet. My goal was not to claim blazars as unique and exotic beasts, but to give them a natural place within the "zoo" of active galaxies.

These ideas came together in a unified model of active galactic

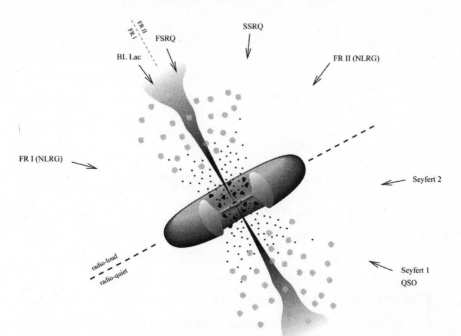

FIGURE 26. In this unified model of active galactic nuclei (AGN), the "zoo" of AGN can be thought of as variations on a basic theme. The power comes from accretion onto a central supermassive black hole, but what observers see depends on their orientation with respect to the inner accretion disk, a larger dusty torus, and the twin relativistic jets. The names of the "zoo" animals we observe are shown around the edge. Types of AGN like Seyfert galaxies, radio galaxies, and blazars are fundamentally the same. This model explains many but not all of the differences between active galaxies. *NASA/ Goddard Space Flight Center/Fermi Gamma-Ray Space Telescope*

nuclei (or AGN). The central idea is that all active galaxies are powered by accretion onto a supermassive black hole and that the observed differences are caused mostly, but not totally, by orientation (Figure 26). Observed properties are strongly affected by obscuration and by the fact that gas in jets is moving close to the speed of light. The intrinsic properties of the nucleus depend on the type of host galaxy, the spin of the black hole, and its accretion rate.[49] Regardless of how differently it presents itself, the elephant is one beast.

4.

GRAVITATIONAL ENGINES

THE DISCOVERY OF active galaxies transformed astronomy. Until then, the universe was thought to be made of stars and gas, gathered by gravity into galaxies, and the galaxies were silently gliding apart as the universe expanded. But learning that the nuclear regions of certain galaxies pump out vast amounts of energy across the entire electromagnetic spectrum changed our understanding of galaxy structure. The discovery also raised questions. How does a supermassive black hole form and grow at the center of a galaxy? What is the evidence that gravity power can create such spectacular phenomena as quasars?

The first answers came from a surprising direction: the center of our own galaxy.

To recap, black holes are gravitational engines. They convert gravitational potential energy into radiant energy. In other words, they use matter to create light. As matter accelerates toward the event horizon it emits high-energy electromagnetic radiation. The efficiency of this process is dozens of times higher than the nuclear fusion that powers stars like the Sun. Ironically, these quintessentially dark astronomical objects can be the brightest for their mass in the universe.

The Big Black Hole Next Door

Zeus was profligate, mating both with goddesses and mortals. His son Hercules was born to a mortal woman, but he let the infant breast feed on the milk of his divine wife, Hera, while she slept. Hera was furious when she woke, and pulled her breast from the infant's lips. The milk spilled across the sky. Hence we call the ragged band of light marking our stellar system the Milky Way, or "galaxy," after the Greek word for milk.[1]

Over 400 years ago, when Galileo pointed his primitive telescope at the gauzy light of the Milky Way, he saw it splintered into a myriad of faint stars. We know now that the Milky Way's patchiness is due to dust, which reddens and dims starlight. The dark patches aren't places where stars are absent; they're places where stars are obscured. Light traveling to us from the center of the galaxy about 27,000 light years away is almost totally blocked.[2] Only one in a trillion photons makes it out. We might as well try to look through a closed door.

Karl Jansky, the father of radio astronomy, showed in 1933 that radio emission from the Milky Way peaked in the constellation Sagittarius—which matched William Herschel's observation that the Sagittarius housed the densest sector in our "city of stars." Radio waves are unaffected by dust. But Jansky's simple radio antenna couldn't pin down the position of the radio emission very accurately. In 1974, Bruce Balick and Robert Brown used the Very Long Baseline Interferometry method to show that the radio source at the center of our galaxy is a very small object.[3] More recent observations reveal it to be the most compact radio source in the sky (Figure 27). This source isn't like the other two sources found in early surveys. Sagittarius A* has similar radio brightness to Virgo A and Cygnus A, but Virgo A (M87) is an active, elliptical galaxy at a distance of 54 million light years, while Cygnus A is a distorted galaxy

750 million light years away. The center of the Milky Way is millions of times less powerful than those two archetypal radio galaxies, so it seems to be a different phenomenon.

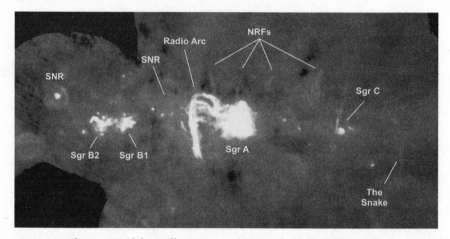

FIGURE 27. The center of the Milky Way is opaque to visible light, but radio waves can travel freely through the galaxy. This radio map shows the region within a few hundred light years of the galactic center; brighter areas represent more intense radio emission. Some of the features identified are non-thermal radio filaments (NRFs) and supernova remnants (SNRs). At the center of the region marked Sgr A is the most compact radio source known, discovered by Karl Jansky in 1932. *F. Yusef-Zadeh/NRAO/AUI/NSF*

What kind of radio source could be so puny? In 1974, the young University of Cambridge theorist Martin Rees hinted at the answer in a paper on black holes that was overlooked at the time.[4] He argued that a massive black hole might be dark because it wasn't accreting any matter, and he was the first to suggest that it might be detected by its influence on the stars orbiting nearby.

It took a while for technology to catch up with this idea. The first problem was the dust between us and the galactic center. Dust particles absorb and scatter light efficiently, but they interact far less with longer-wavelength photons. By shifting our attention from visible light at 0.5 microns to the near infrared spectrum at 2 microns, the dimming toward the galactic center drops from a factor of a trillion to a factor of 20. That's like looking through smoky glass instead of a closed door. Infrared detectors first emerged from physics labs in the 1970s, but they had just a single element or "pixel," so making an image meant tediously scanning the telescope in a grid pattern. Like an Italian sports car, the detectors were expensive, temperamental, and prone to break down. By the mid-1990s, the first megapixel arrays were being used, and digital infrared astronomy matured the way optical astronomy had fifteen years earlier.[5]

The second problem was the high density of stars, which caused the images to overlap and bleed into one another.[6] Let's visualize the physical situation. There are 10 million stars within a few light years of the center of the Milky Way. That's a density 50 million times higher than the neighborhood of the Sun. If we lived there, the night sky would be spectacular. The light of a million stars would shine hundreds of times brighter than the full Moon; you could read a newspaper by starlight alone. On the other hand, it would be almost impossible to do optical astronomy in such an environment. Even worse, life on any planet would be challenged. Supernova explosions would be frequent, and possibly near enough to devastate biology. Frequent interaction of the stars would interfere with solar systems, causing planets to be tossed into deep space. Comet clouds at the periphery of solar systems would be disrupted, leading to impacts and mass extinctions far more often than they occur on the Earth. We should be grateful to be located in a quiet suburb of the Milky Way.

The convergence of two technologies—infrared detector arrays and techniques to sharpen astronomical images—suggested an exciting experiment. Make the sharpest possible infrared images of the galactic center. Find stars within a few light years of the compact radio source that move fast enough that their motions can be tracked from year to year. Then use their orbits to deduce the mass in the central region of our galaxy.

A research group at the Max Planck Institute near Munich was the first to attempt it. They used a 3.5-meter telescope in Chile designed specifically to make crisp images. A couple of years later, a group at UCLA began the same experiment using their newly built 10-meter Keck telescope in Hawaii, the world's largest. Both groups had to fight the blurring of images by the Earth's atmosphere. If you look at a star with a telescope at an excellent astronomical site, you'll see a bright core of light randomly dancing and jittering around, surrounded by evanescent speckles of light. The speckles are caused by rapid variations in air density and temperature in the Earth's upper atmosphere, which bend light rays, blurring and jumbling the image. A long-exposure image will average the speckles, making the star appear smooth but blurred. Short-exposure images "freeze" the atmosphere. Researchers can process, shift, and layer these images to create a much sharper image; however, the method is very tedious. Thousands of images, each exposed for a few tenths of a second, have to be analyzed and combined to make a single sharp image.

After several years of following this painstaking method, the researchers isolated several dozen stars in the crowded region, which were then tracked in their elliptical orbits (Figure 28). Each star contributes to an estimate of the mass that's driving their collective motions.[7]

Both groups of researchers reached the same stunning conclusion: some of the stars close to the galactic center were moving faster than 300,000 mph, and the implied mass in the central few light years of the galaxy was millions of times the mass of the Sun. But nothing like the corresponding amount of starlight was coming from that region. Even the hypothesis of a dense cluster of dim stars failed by orders of magnitude to account for the large central concentration of mass. The evidence pointed in just one direction: a single, compact, dark object millions of times the mass of the Sun. There was a supermassive black hole on our doorstep.

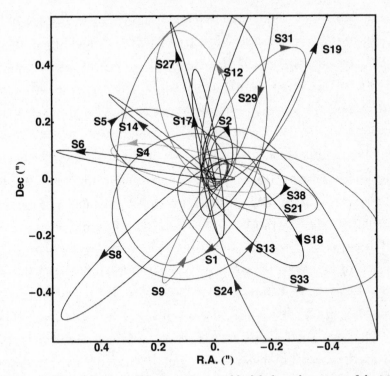

FIGURE 28. Stellar orbits around the supermassive black hole at the center of the Milky Way, measured over sixteen years using infrared imaging with adaptive optics to make the images extremely sharp. At a distance of 27,000 light years, one arc second, which is the width of the image, is 0.1 pc or 4 light months. This data can be used to derive a Keplerian orbit for each star and the resulting mass of the black hole is measured very reliably. *S. Gillessen/Institute of Physics/ApJ 692, 2009/MPE Galactic Center Team, reproduced with permission/copyright AAS*

Stars at the Edge of the Abyss

As Robert Frost wrote, "We dance around in a ring and suppose, but the Secret sits in the middle and knows." Nature guards its secrets zealously, and it takes grit and determination to shed light on them. The quest to demonstrate that the Milky Way harbors a massive black hole featured one of the more intense rivalries in astronomy.

On one side was Reinhard Genzel. A burly, red-haired, mustachioed man with a ruddy complexion, Genzel was the director of the Max Planck Institute for Extraterrestrial Physics in Garching, Germany. In the pantheon of plum jobs in astronomy, Genzel had one of the best. Directors of the various Max Planck Institutes are appointed by Germany's elite scientific organization, the Max Planck Society, and they hold their jobs for life. They wield top-down and absolute power and can bring the resources of a large organization to bear on the research questions of their choice. Genzel was appointed director when he was just thirty-four, and his group was the first to publish results on the galactic center and claim the presence of a dark, compact mass.

On the other side was Andrea Ghez. A New Yorker of Italian ancestry, she was four years old when she announced to her mother that she wanted to be the first woman to set foot on the Moon. She became an astronomer instead, earning degrees from MIT and Caltech. She was just twenty-nine, and an assistant professor at UCLA, when she observed the galactic center for the first time, using the Keck telescope in Hawaii. She went back to Keck the following year and saw that the stars had moved in a short time: "If there was a black hole, these things should move quite a lot. And that first year, we could see very easily that the stars had moved, and we were just thrilled. I think the thrill was heightened by instrument failure at the beginning of our night. It's very difficult to get Keck time—you might only get a few nights a year. . . . Just before the center of the galaxy set and we wouldn't be able to see it any more, things came together and we got the picture."[8]

Sometimes in astronomy there's a moment of discovery, and other times data has to be accumulated painstakingly over years, ascending only slowly to the level of definitive proof. In this case, both research groups, led respectively by a scientist at the peak of his powers[9] and a rapidly rising female star,[10] knew where to look for their discovery and both knew exactly what they were looking for. Success would take perseverance and careful experimental technique.

In the early 2000s the German group moved up from a 3.5-meter telescope to the 8.2-meter Very Large Telescope (VLT), run by the European Southern Observatory in Chile. In the mid-2000s, both groups began using adaptive optics,[11] a huge innovation that transformed modern astronomy, allowing astronomers to "cheat the atmosphere" and make images as sharp as the diffraction limit of the large telescope they're using. With this technique, the image blurring and distortions created by the atmosphere are rapidly compensated for by a flexible secondary mirror. Light from a powerful laser is bounced off the upper atmosphere where turbulent motions occur, small deviations in the wave front of light are measured hundreds of times a second, and corrections feed into mechanical actuators attached to the back of the secondary mirror.

Adaptive optics allowed scientists to apply Kepler's laws to stars observed at the center of the galaxy, which are in motion like an angry swarm of bees. One star has been tracked over its entire sixteen-year orbit.[12] Astronomers have watched a star or a gas cloud get torn apart as it plunged toward the zone of strong gravity.[13] The material eaten by the black hole probably led to a sequence of X-ray flares in 2014. Using Kepler's laws, scientists can derive the mass of the object causing their motion. The American and German groups battled to an honorable draw. Meanwhile, radio astronomers showed that the radio source was as small as the expected size of the event horizon.[14] The mass is measured as 4.02 million times the mass of the Sun, with an error of only 4%.[15] Because these calculations are now possible, researchers no longer need to qualify their writing with adjectives like "candidate" and "hypothetical." The existence of a supermassive black hole has been proven beyond any reasonable doubt.

The Dark Core in Every Galaxy

Quasars are extremely rare—a million times rarer than normal galaxies.[16] On average, you'd need to search a volume of space a billion light years on a side to find one. As soon as active galaxies were discovered, astronomers wondered whether every galaxy goes through an active phase. A bright young theorist from England had an important insight.

Donald Lynden-Bell was eclectic in his interests. He worked on fluid dynamics, ellipsoidal orbits in galaxies, negative heat capacity, and a gravitational effect called violent relaxation before turning his attention to quasars. In a prescient paper in 1969, Lynden-Bell deduced that quasars have active periods and are rarely very bright. He estimated that dead quasars should be common and that the nearest might be less than 10 million light years away (just four times the distance to the Andromeda galaxy). He argued that these dark central masses would gather many stars around them and be detectable by their influence on those stars.[17]

I was twelve when Lynden-Bell wrote his paper so I didn't read it until much later. But he influenced my life around that time. My father and I were on a road trip in the south of England. We visited relatives in Hastings, sat on the shingle beach at Brighton, then headed across the South Downs to Herstmonceux Castle. The castle was almost perfect—medieval, constructed of red bricks, and surrounded by a moat. But I was too old for castles. Casting around for something else to do, my father spotted a sign for the Royal Greenwich Observatory, which occupied part of the castle. There was a talk starting in half an hour.

Donald Lynden-Bell paced beside the lectern, head down, deep in thought. We took our seats—and soon realized we were in way over our heads. Lynden-Bell punctuated his lecture with expansive hand gestures, and moments when he turned to the blackboard and scribbled a blizzard of equations. The talk was about galaxies and black holes. Apart from the broadest of the brushstrokes, the lecture was inscrutable.

I had no idea of my future path, and at times thought I might become a farmer, or an architect, or a pilot. But something about the tweedy theorist struck a chord with me. He told me there were myriad galaxies out in space waiting to be measured. He said that these galaxies harbored dark objects that could be understood with beautiful mathematics. He projected an infectious excitement that the universe was knowable. And so a small seed was sown.

Lynden-Bell was suggesting that all massive galaxies have supermassive black holes at their core, and that the reason quasars are rare is that they spend only a small fraction of their lives actively accreting gas. We see just the small fraction that are "switched on." Most are hibernating, with no "food" nearby, and their pulse and life signs dialed down to very low levels.

How do you find something compact, massive, and dark at the center of a galaxy? It depends on being able to isolate a central region where the black hole dominates the gravity—the sphere of gravitational influence. Within the radius of this sphere, motions of stars and gas are driven by the black hole. Beyond this radius, motions are driven mainly by stars near the center of the galaxy; the black hole is a minor contributor. For a big galaxy containing a beefy black hole 100 million times the mass of the Sun, this distance is about 10 parsecs or 33 light years.[18] That's extremely close to the center of a galaxy 100,000 light years in diameter; if the galaxy was the size of a dinner plate, the region dominated by the black hole would be the size of a mote of dust. In a remote galaxy, it's very difficult to observe star or gas motions on this tiny scale.[19]

We can measure that the black hole at the center of our galaxy is only 27,000 light years away, 100 times closer than the center of the nearest large galaxy, Andromeda. Astronomers are able to sample stars on a scale 1,000 times smaller than the gravitational sphere of influence, giving them an excellent handle on the mass of the black hole. This makes it the "gold standard" for massive black hole detection—verifying the existence of one beyond a reasonable doubt. But scientists were also hungry to bag hibernating black holes in galaxies other than our own. They pinned their hopes on the Hubble Space Telescope.

When the telescope was first launched in 1990, it was a bitter disappointment. It was built to take super-sharp images from Earth orbit, and was designed to make images up to 10 times sharper than even the best ground-based telescopes that had come before it. But when the Hubble Space Telescope's first images came back, NASA officials were puzzled—and then mortified. An error in final testing in the lab had resulted in the primary mirror having spherical aberration, giving distorted images. The problem was misunderstood by the media, which accused Hubble of having a cheap, crappy mirror. In fact, it was the most precisely machined mirror in history, ground to a precisely wrong shape because a calibration lens had been positioned incorrectly in the lab test. It took three years and a high-risk Shuttle mission involving thirty-five hours of astronaut space walks to fix the problem.[20] With the telescope restored to full health, it was able to take crisp images of galaxy nuclei tens of millions of light years away.[21]

To search for black holes in nearby galaxies, the telescope is pointed so that the nucleus of the galaxy falls within the narrow slit of the spectrograph. Spectra can be extracted at different positions along the slit, corresponding to different distances from the center of the galaxy. The width of spectral features gives the average velocity of material—gas measured using emission lines if the galaxy is a spiral, stars measured using absorption lines if the galaxy is an elliptical.[22] The telltale indication of a black hole is a sharp increase in the spread of gas or star velocities very close to the center of the galaxy (Figure 29). The nearest big concentration of galaxies is the Virgo Cluster, 60 million light years away. For a galaxy in the Virgo Cluster, the angular size of the sphere of gravitational influence is 0.14 arc seconds. That's barely twice the angular resolution of the Hubble spectrograph, so looking for black holes at these distances pushed the space telescope to its limits.

A decade of this slow and difficult work produced success: about two dozen black holes were detected in nearby galaxies.[23] Our immediate neighbor, the Andromeda galaxy (M31), has a 100-million-solar-mass black hole, surrounded by a cluster of young blue stars. We're not yet certain how they can form and survive in such an extreme environ-

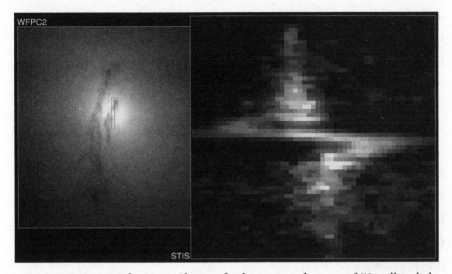

FIGURE 29. M84 is in the Virgo Cluster of galaxies, at a distance of 50 million light years. The image on the left shows the central region of the galaxy, which is crossed by dust lanes. The rectangle shows where the slit of a spectrograph on the Hubble Space Telescope was placed to take the data shown on the right. The zigzag shows the gas velocities measured along the slit, with larger horizontal displacements indicating larger velocities. If there were no black hole in M84, the trace would not have very large velocities near the center of the galaxy. *G. Bower, R. Green/NOAO/NASA*

ment,[24] although this may be a general phenomenon in spiral galaxies. Andromeda's dwarf companion, M32, also has a black hole, slightly less massive than the Milky Way's, weighing in at 3.4 million solar masses within a region smaller than one light year.[25] At the other end of the size scale is radio source Virgo A, now known as the giant elliptical galaxy M87. The black hole at the center of M87 is a true monster, 6.4 billion times the mass of the Sun.[26] Its event horizon is larger than the Solar System! Black holes in the local universe vary greatly in size, spanning a factor of 2,000 in mass.

Forty years after he wrote his prescient paper, Lynden-Bell stood onstage in Oslo to accept the inaugural Kavli Prize. Fittingly, the man beside him was Maarten Schmidt, the discoverer of quasars. Lynden-Bell's insight about black holes was a perfect complement to Schmidt's contribution: darkness lurks in the heart of every galaxy.

Baron Rees of Ludlow Tames the Beast

It had taken just a decade for black holes to go from being an esoteric theoretical concept to being the centerpiece of massive star evolution and the explanation for activity in the nuclei of galaxies. Cambridge University was the place to be if you were a theorist. Donald Lynden-Bell got his PhD there in 1961 and wrote his seminal paper on dead quasars in 1969. Stephen Hawking got his PhD there in 1966 and wrote his paper on radiation from black holes in 1974. Martin Rees got his PhD a year after Hawking, and wrote his influential paper about supermassive black holes also in 1974.

It was Martin Rees—not yet a lord—who put the role of supermassive black holes as gravity engines on a firm theoretical footing. To any student of cosmology, Rees is a titan. His awards are legion: the Templeton Prize, the Dirac Medal, the Newton Prize, the Bruce Medal, the Descartes Prize, the Japanese Order of the Rising Sun, and the British Order of Merit. He has been President of the Royal Society, Master of Trinity College, Cambridge, director of the Institute of Astronomy, Plumian Professor of Astronomy and Experimental Philosophy at Cambridge University, and England's Astronomer Royal. (Typically modest about his professional duties, Rees characterized the obligations of this last post as "so exiguous one could perform them posthumously.")

So when I first met him I was expecting someone larger than life. In the flesh, he's a short man with a hawklike nose and piercing gray eyes. He speaks so softly you have to lean in to hear him. His voice has the lilting cadence of Shropshire, where he grew up (Figure 30). Rees showed that accretion onto a spinning black hole can lead to twin relativistic jets and nonthermal emission across the electromagnetic spectrum, from meter-length radio waves to gamma rays with a wavelength smaller than a proton.[27]

Active galaxies are a multi-scale phenomenon, meaning that a range

FIGURE 30. Martin Rees has been one of the world's foremost theorists for over forty years. He was one of the first to understand how black holes act as gravitational engines to generate huge amounts of energy and drive relativistic jets of plasma. Rees is a Fellow of Trinity College, Cambridge, and Emeritus Professor of Cosmology and Astrophysics. He has been both Astronomer Royal and President of the Royal Society. In 2005, he was appointed to the House of Lords. *M. Rees/University of Cambridge*

of wavelengths and methods are needed to understand them. They are fascinating to theorists but difficult to observe. Diffuse lobes of radio emission can extend several million light years from a galaxy. The fueling of the central, massive black hole is governed by the host galaxy's nearby environment and gas content. On a scale of several hundred light years, there's a nuclear star formation region and a dusty torus. In the middle of the dusty torus, on a scale of light weeks to light months, dense and fast-moving gas clouds produce broad emission lines. Even closer in, the hot accretion disk pumps out a large amount of ultraviolet and X-ray emission on a Solar System scale. This emission is smoothly distributed across wavelength as a continuum. Finally, at the center of these nested Matryoshka dolls, the supermassive black hole exerts its gravitational reach over a factor of billions in scale.[28]

Rees is partially responsible for black hole accretion becoming the rarely questioned paradigm of how active galaxies work. At a conference in 1977, astrophysicist Richard McCray lampooned the tendency of astronomers (and scientists of any stripe) to be in thrall to a popular idea. He showed a diagram with simple stick figures and boundaries drawn as dotted lines, one representing the sphere of gravitational influence, and the other, the event horizon of a black hole (Figure 31). Let's hear his description of the cartoon and the sociology behind it: "The system is characterized by two radii. Beyond the accretion radius, astrophysicists are sufficiently busy with other concerns not to be significantly influenced by this fashionable idea. But others within this radius begin a headlong plunge toward it. There's little communication among individuals as they follow random ballistic trajectories, depending on their initial conditions. In their rush to be the first, they almost always miss the central point, and fly off on some tangent. With a sufficient number of astrophysicists in the vicinity of the idea, communication must occur, but it usually does so in violent collisions. . . . The only lasting effect is that some individuals may have crossed the rationality horizon, beyond which the fashionable idea has become an article of

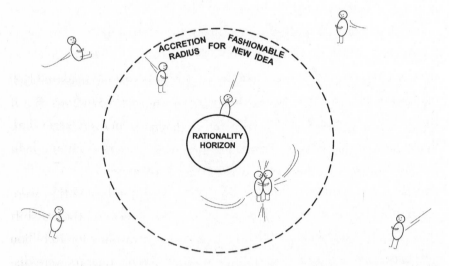

FIGURE 31. A satirical view of how astrophysicists react to a fashionable idea like black holes. Some fly past without getting trapped. Others collide, generating heat but not much light. Most are trapped within the sphere of influence of the idea, and some slip toward the "rationality horizon" where they lose their healthy skepticism. *R. McCray/ University of Colorado*

faith. These unfortunate souls never escape."[29] McCray's presentation was tongue-in-cheek. He was convinced by the black hole argument, but was reminding his colleagues not to leave skepticism at the door.

Recall that by the mid-1980s only the Milky Way showed compelling evidence for a massive black hole. In one of his papers, Rees included a flow chart, demonstrating the way in which gas from the intergalactic medium drizzles into galaxies and slowly finds its way to the nuclear regions. This gas, and gas jettisoned from evolved stars, feeds the formation of a nuclear star cluster, which is a dense aggregation of many thousands of stars brought together by gravity. The star cluster can't sustain itself against the gravity of so many stars, so it collapses into a large black hole and the black hole grows by devouring gas and stars. Though he presented it as a flow chart, Rees had sound physical arguments for each step. The result gave massive black holes an air of inevitability. This is a gift of the best scientists: to take a complex argument and make it seem obvious.

Using Quasars to Probe the Universe

So far we have been focused inward, trying to understand massive black holes by looking at the effects they have on their surroundings. But it turns out that black holes can be used to diagnose an even larger dark ingredient of the universe. The technique uses quasars as intense light sources that can be seen across vast distances of space.

When quasars were discovered, their redshifts indicated they were at very large distances. Two years after their discovery, the redshift record was $z = 2$, indicating light that had been traveling for 10 billion years, or 75% of the age of the universe. At the time quasars were discovered, the redshift record for a normal galaxy was only $z = 0.4$, indicating light that had been traveling for 33% of the age of the universe. The use of supermassive black holes as distant light beacons opened up a new field of astronomy.

Imagine a very long black box that's dark inside but has open ends. Shining narrow beams of light through the box and detecting them at the other end would reveal whether or not there was anything in the box. An obstruction would block the light completely and even something nebulous like gas would dim the light. When astronomers used spectroscopy to spread out the wavelengths of quasar light finely, they saw that the smooth light distribution was peppered with "notches" where the light was missing or absorbed. The importance of this absorption was first realized 200 years ago, when Joseph von Fraunhofer mapped dark, narrow lines in the Sun's spectrum and Gustav Kirchhoff showed the lines were caused by chemical elements in the Sun's cool outer atmosphere.

Quasar spectra have two types of absorption lines.[30] Absorption lines are narrow, dark regions in the spectrum where light has been absorbed by objects in the intervening space. There are lines due to elements created in stars, such as neon, carbon, magnesium, and silicon. There's also a thicket of hydrogen absorption lines at short wavelengths. After much investigation, it became clear that the first type of line is caused by chemically enriched gas in galaxy haloes along the line of sight to the quasar. The hydrogen lines are due to primordial hydrogen in the vast spaces between galaxies (Figure 32).[31]

Absorption line spectroscopy is sensitive to tiny amounts of gas, so dim or dark gas clouds only 10 to 100 times the mass of the Sun can be detected at distances of billions of light years. The expanding universe model gives the relationship between redshift and distance, so a spectrum, which is a map of wavelengths, is easily converted into a map of redshift or distance. Returning to the earlier analogy, the long black box is a path through the universe, quasars are flashlights at the far end, and astronomers take spectra of those beams of light to diagnose the intervening material. Think of it as a core sample through the universe, mapping the material across cosmic time rather than geological time. Since quasars have been found with redshifts as high as $z = 7$, the samples can encompass 95% of the age of the universe. Quasar absorption spectra

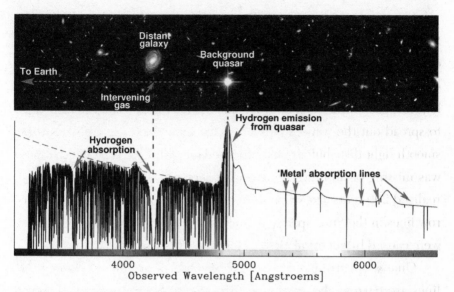

FIGURE 32. Quasars are at very large distances and they act like flashlights to illuminate intervening material that may be dark and difficult to detect otherwise. Above: Quasar light passes through a large galaxy and its halo and numerous small clouds of hydrogen in the intergalactic medium. Below: The spectrum of the quasar is a graph of intensity versus wavelength. The large galaxy imprints absorption lines from heavy elements in the red part of the spectrum, while small hydrogen clouds imprint a "forest" of narrow absorption lines in the blue part of the spectrum. *M. Murphy/Swinburne University*

have been used to show that there's 8 times more material in the space between galaxies than in all the stars in all the galaxies in the universe.[32]

Quasars are used as probes of the universe in another way. Let's return to light shining through the "long black box" of the universe. Space is mostly empty, but there's a small chance that light from a distant quasar will pass directly through a galaxy or a cluster of galaxies. Einstein's theory of general relativity says that light will be deflected by the mass of an intervening object. If the alignment is perfect, the quasar point source is turned into a circle of light called an Einstein ring. If the alignment is slightly off, the point source is seen as twin images.[33] The odds of this happening are only 1%, so the phenomenon wasn't noticed until hundreds of quasars had been discovered. Since lensing is sensitive to matter that's dark as well as to visible matter, it has been used to show

that dark matter is a ubiquitous component of galaxies, outweighing normal matter by a factor of 6.

It's an unexpected bonus that quasars are such excellent probes of
the universe. The universe contains 10,000 billion billion stars in several hundred billion galaxies. Yet quasars have told us that there's much
more mass in the spaces between galaxies and even more mass that's
dark and undetectable by any other means. All those stars and all those
galaxies are just 2% of the material universe!

Weighing Black Holes by the Thousand

Let's continue the story of quasar discovery. Spectroscopy is required
to reveal and to understand quasars. An optical spectrum is used to
measure a redshift, which can be used to calculate luminosity. A high-
quality spectrum can be used to measure black hole mass. But progress
was slow. Large telescopes could only take spectra of one candidate at a
time. Through the 1960s and 1970s, the number of known quasars crept
up from a few dozen to a few hundred.

The first breakthrough involved telescopes with special optics to
make images of large swathes of sky. The wide-field Schmidt Telescope at
the Palomar Observatory was completed in 1948, and during the 1950s
it was used to survey the entire northern sky in two colors, with nearly
2,000 photographic plates. Each plate covered 36 square degrees, about
the size of your closed fist held at arm's length. The survey was funded
by the National Geographic Society, as a cosmic extension of their goal of
mapping the world. A twin of the Palomar Schmidt was built in Australia
and it surveyed the southern sky during the 1970s. Each image included
a million galaxies and 10,000 quasars and active galaxies.

Finding the 1% of galaxies with nuclear activity required additional
information. Optical designers developed a large prism that could be
placed in the optical path of a Schmidt telescope. The prism smeared

each faint source of light into a tiny spectrum on the photographic plate. Quasars have strong and broad emission lines; it was hoped that they would stand out because the emission line would appear as a blob on top of the streak (Figure 33). It required great expertise to find quasars by eye, but machines were developed to scan and digitize the plates and look for quasars using algorithms to sift them from the much more numerous stars and galaxies.

FIGURE 33. Multiple spectra can be taken by putting a large prism in the optical path of a telescope, so that there is a direct image of each object (the dots), and a spectrum to the right (the horizontal streaks). Most of the objects in any field of view are stars or galaxies with smooth and featureless spectra at this low resolution, but rare quasars stand out because they have strong, broad emission lines that look like blobs on top of the streak. The quasar 3C 273 is near the center of this photographic image. *David Haworth*

I got my feet wet with quasar-hunting using this method. The place was Coonabarabran in the Warrumbungle Mountains of New South Wales, Australia. That sleepy town on the edge of the Outback is the home base of the U.K. Schmidt Telescope, the southern hemisphere

twin of the Palomar Schmidt.[34] I was sent there as a graduate student
at Edinburgh University to help with a photographic prism survey. It
wasn't exactly a hardship to go from the gloomy Scottish winter to the
hot Australian summer. Within a few days of arriving I'd been trained
in the darkroom and I was observing all night and developing the photographic plates before going to bed. The plates were 14 inches on a
side, a millimeter thick, and very tricky to handle in the dark. After all
this time, it still pains me to admit that I broke more than a few, wasting hours of telescope time. Occasionally I was literally pained by the
razor-sharp edges, adding drops of my blood to the developer or the
fixer fluid.

When the sky was clear and the plate was well exposed, it was worth
the effort. Each plate was a negative with a pale gray background and
thousands of small dark streaks representing the spectra. I'd sleep until
lunch and in the afternoon mount the plates in a light box and scan them
with a microscope. My elusive quarry was a streak with a blob at the
blue end, looking like a tadpole. The blob was the hydrogen emission
line that would distinguish a quasar from a hot star. I remember the
jolt of excitement I felt when I found my first quasar, and it didn't fade
after finding dozens, though my vision started to blur after hours staring through the microscope. Each of these little tadpoles was a massive
black hole billions of light years away, pouring a torrent of radiation into
the universe. After bagging my hundredth quasar I went for a celebra-
tory hike in the local mountains, bushwhacking through wild terrain.
At dinner, the local astronomers gave me a royal ribbing, reminding me
that Australia has three of the five most venomous spiders in the world,
and four of the five most venomous snakes.

The second breakthrough took place in the 1990s, when photographic plates were replaced with large-format electronic detectors, or
CCDs. Using fibers or slits, the light from hundreds of targets is gathered and projected onto the CCD. Large telescopes now have spectro-
graphs that can cover a square degree or more—several times the area

of the full Moon. The preeminent quasar-hunting tool is the 2.5-meter telescope doing the Sloan Digital Sky Survey. The telescope wouldn't make a list of the top fifty largest telescopes in the world (and of course, the Hubble Space Telescope wouldn't either), but its exquisite spectrograph and CCDs give it an extraordinary grasp of light. It has measured redshifts for 2 million galaxies and 500,000 quasars (Figure 34). Cru-

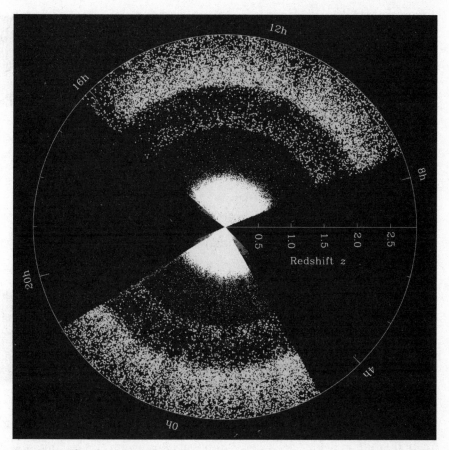

FIGURE 34. The Sloan Digital Sky Survey used a 2.5-meter telescope in New Mexico and an efficient multiobject spectrograph to measure the redshifts of an unprecedented number of galaxies and quasars. The BOSS (Baryon Oscillation Spectroscopic Survey) project observed 500,000 galaxies and 100,000 quasars, a small fraction of which are shown in this "pie chart" of the sky. Redshift or distance increases radially. Galaxies are dots at redshifts less than 1, and quasars are dots from redshift 1.5 out to 3. *M. Blanton/Sloan Digital Sky Survey*

cially, those digital spectra are much better than the little streaks I used in the 1970s to discover quasars. Spectra from the Sloan Survey are high enough quality to allow a measurement of a black hole's mass.

We've seen how difficult it is to "weigh" a supermassive black hole. The closest one, at the center of our own galaxy, was weighed accurately using individual stars that loop around it on elliptical orbits. A second precise measurement of black hole mass was made in 1995, when radio astronomers discovered water masers— long-wavelength versions of lasers, produced when conditions naturally occur for a gas (in this case, water molecules) to emit intense and pure radiation—orbiting in a thin disk at the center of the nearby active galaxy NGC 4258. Other galaxies, too, show maser emission from water molecules in their dense nuclear regions, and the resulting spectral lines allow the masers' velocities to be very accurately measured by radio methods.[35] In NGC 4258, the positions and velocities of the masers match Kepler's laws of motion, implying a central mass of 3.82 million times the mass of the Sun with an error of only 0.3%. Maser emission extends to less than a light year from the center of the galaxy, or 1,000 times smaller than the sphere of gravitational influence, so large mass is concentrated in a region that would normally only contain a few hundred stars. A black hole is the only viable interpretation. Maser emission is rare, so this observation proved hard to replicate, but it may soon be possible using interferometry at millimeter wavelengths.[36]

The quiescent black holes of galaxies in our cosmic neighborhood can be weighed using the motion of gas or stars near the nucleus, but several decades of this work have yielded masses for just seventy black holes. Extending these measurements beyond the Virgo Cluster, which is about 60 million light years away, is impossible using current techniques.

Quasars, as we've seen, have supermassive black holes that act as gravity engines by converting the mass that falls into intense radiation. Why not use the brightness to infer the black hole mass? It's a good idea, but it doesn't work in practice. Far from being flashlights with a standard brightness, the brightness of quasars ranges by a factor of thousands from one quasar to the next. For a particular black hole mass, the brightness depends on the efficiency of accretion, the black hole spin rate, and the amount of gas and dust in the central regions. Unfortunately, quasar power is a poor guide to black hole mass.

Just when it seemed as if astronomers had reached the end of the road, they came up with a clever method to infer black hole mass in nearby active galaxies. It uses one of the signature features of quasars: their broad emission lines. The hot gas producing these emission lines is within a light year of the central object, so its motion is dominated by the black hole. The gas in this region should obey a simple equation, $M_{BH} \approx RV^2/G$, where G is the gravitational constant, and V is the velocity of the gas. The same equation could give us the mass of the Sun if we knew the velocity and distance of an orbiting planet. In the case of a black hole, the velocity of the orbiting gas is easily derived from the width of the emission lines. That just leaves R, the size of the region producing the broad emission lines, as an unknown quantity. Various physical arguments suggest the region is about 0.01 parsecs, or 10 light days across—10 times larger than the Solar System (Figure 35).[37] This is far too small to be resolved by any telescope, for most galaxies. So how can it be measured? The clever method makes use of the fact that the light intensity from quasars and active galaxies varies with time.

Let's visualize the situation. The accretion disk that generates the enormous brightness of a quasar is so small that we can consider it a point source of light. The brightness varies on timescales of days, which was one of the original arguments for supermassive black holes, since the source could be no bigger than the light travel time across it. The logic of this argument is that if the light variations arise from a single object, more rapid variations imply a smaller object. Light from the central point source travels out and hits the fast-moving gas that causes emission lines. This gas responds or "reverberates" to the varying point source with a delay, t, given by the light travel time across the gas, $t = R/c$, where c is the speed of light. It's called reverberation

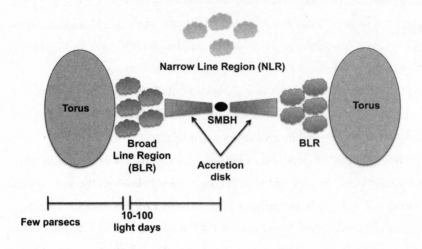

FIGURE 35. A schematic cross section of the inner region of an active galaxy or quasar shows the ingredients that can be used to measure the mass of the central supermassive black hole (SMBH). When light from the central "engine" varies, gas clouds in the broad line region respond with a time delay of 10 to 100 light days. The velocity of those clouds is measured by the width of their spectral lines. As a result, the mass of the black hole that drives the motions can be estimated. The technique is called reverberation mapping. *C. Ricci/Catholic University of Chile*

mapping, because we're mapping the way light from the point source results in "echoes" from the gas. The time it takes the echoes to arrive gives the size of the region of hot gas.

The required observations are simple but tedious. An observing "campaign" is set up, with telescopes around the world measuring spectra for a sample of quasars or active galaxies. Having a handful of telescopes around the world gives twenty-four-hour coverage of the variation and ensures data even if one or two sites are clouded out. Spectra are gathered over weeklong observing runs scattered through the year, so all timescales from days to months are sampled. The emission-line gas "responds" to the radiation from the black hole with a time delay due to light travel time. The time delay gives the size of the broad line region, which in turn gives the mass of the black hole.[38]

Thus, reverberation mapping relies on time resolution rather than spatial resolution. The method was first applied to NGC 5548, one of Seyfert's original active galaxies; its central black hole is 65 million times the mass of the Sun, with an uncertainty of 4%.[39] Intensive monitoring campaigns with small telescopes have yielded sixty black hole masses for nearby active galaxies.[40] The research shows that more powerful active galaxies have larger regions of fast-moving gas.

This is where it gets fun. The painstaking reverberation mapping work shows how the size of the emission region relates to the luminosity of the active galaxy. Now, rather than carrying out a long-term monitoring campaign involving hundreds or thousands of measurements for an active galaxy of interest, a single spectrum can be used for an estimate of the black hole mass. The emission line width gives V and the luminosity gives R, which is all that's needed in the equation $M_{BH} \approx RV^2/G$. Black hole masses from single spectra are uncertain by a factor of 3, or 300%, which isn't great, but it's adequate for statistical work. Rather than spending months of observation to acquire a single black hole mass, you can bag 100 black hole masses in a single night. Tens of thousands of masses have been published.[41] Astronomers are harvesting massive black holes on an industrial scale.

Accretion Power in the Cosmos

Matter falls onto a black hole and it heats up. Also, the rotation energy of the spinning black hole accelerates particles which then emit radiation. This process is extremely efficient. If we define efficiency as the energy output divided by the mass-energy of all the input ingredients, accretion onto a black hole is about 10% efficient, compared to 1% for nuclear fission or fusion, and $10^{-7}\%$ for chemical energy. Matter can liberate 10% of its mass-energy as photons just by falling!

How much mass does it take to turn a supermassive black hole into a quasar? Not very much. For a black hole of 100 million solar masses to generate a quasar-like power of 10^{39} watts at an efficiency of 10%, only one solar mass per year has to be accreted.[42] Think of it—snacking on just one star per year can keep a black hole shining brighter than an entire galaxy of stars. As John Updike said, "There is still enough energy in one overlooked star to power all the heavens madmen have ever proposed."[43] But feeding a black hole is a challenge, because the radiation produced by a quasar exerts a pressure that drives matter away from the central source. It's analogous to the phenomenon of radiation pressure making a comet tail point away from the Sun. The inward gravity force of a supermassive black hole must exceed the outward radiation pressure for matter to be accreted.

It took a long time for astronomers to get a complete picture of accretion power in active galaxies. That's because physical processes close to black holes spread energy across an enormous range of wavelengths.[44] For example, the archetypal quasar 3C 273 has been detected at frequencies ranging from 10^8 Hz to 10^{24} Hz—wavelengths ranging over a factor of 10,000 trillion, from radio waves 3 meters long to gamma rays one-third the size of a proton (Figure 36). However, the only wavelengths from that vast range that can be detected at ground-based observatories are a broad swath of radio waves and a narrow slice from near-infrared through optical wavelengths. The rest require specialized satellites in Earth orbit.

Viewing the universe using just one part of the electromagnetic spectrum leads to incomplete information: the elephant problem. Full accounting of accretion power means we have to consider the whole elephant. The radio emission that first drew attention to active galaxies in the 1950s turns out to be a very small fraction of the total power of a quasar. This emission is from relativistic electrons near the black hole and in twin jets. Let's call it the tail of the elephant. The next most important contribution is high energy X-ray emission, which also comes from relativistic electrons. Let's call that the elephant's trunk.

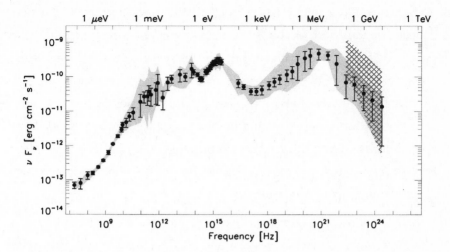

FIGURE 36. The energy distribution of the bright quasar 3C 273 across the entire electromagnetic spectrum, from long radio waves (10^8 Hz) to high-energy gamma rays (10^{24} Hz). The vertical axis is energy and the horizontal axis is frequency (along the bottom) or energy (along the top). Stars and normal galaxies only emit a narrow range of optical and infrared wavelength, so such a broad range of energy indicates gravity energy and particle acceleration close to a supermassive black hole. *S. Soldi/University of Paris/CNRS*

Even more important is infrared radiation from cool dust farther away from the black hole, at temperatures ranging from 10 to 100 Kelvin. The dust is the elephant's leg. The dominant contributor to quasar power is the accretion disk, very close to the black hole. It has a temperature of roughly 100,000 Kelvin and puts out most of its energy at ultraviolet and X-ray wavelengths.[45] This is the bulk of the elephant, its body.

Active galaxies were first discovered by their radio emission, because they stand out in the generally quiet radio sky. But within a few years, astronomers realized that most active galaxies have such weak radio emission that they're invisible to radio searches. Ten times as many are found in optical surveys as in radio surveys. Then, in the 1980s, X-ray astronomers were puzzling over a weak X-ray signal that could be seen all over the sky.[46] They assumed it was the sum of many distant sources that were too weak to be detected individually. But, when they added up the expected X-ray radiation from existing optical samples of active galaxies, it fell short of the X-ray background by a factor of 10. The puzzle isn't completely solved, but it's now clear that the X-ray background is due to active galaxies that are missing from optical surveys.[47] They've been rendered invisible by dust. The presence of dust can radically change the energy distribution of an active galaxy by reprocessing optical emission into infrared emission. Dust doesn't affect X-ray photons, so the clearest and most complete view of the population of active galaxies comes from X-ray surveys.

Massive Black Holes Are Not Scary

Let's alleviate the fear factor associated with black holes. Black holes are not like cosmic vacuum cleaners, sucking in everything around them. Black holes do have a sphere of gravitational influence, like any object with mass, but if the Sun were to suddenly condense to a black hole, the gravity at the distance of the Earth would remain unchanged and the Earth would continue unperturbed in its orbit (although humans would be *very* perturbed by the loss of the Sun's light and energy eight minutes later). Second, we're not in imminent danger of encountering a black hole. A tiny fraction of stars die as black holes, and there are no black holes in the neighborhood of the Sun.[48]

The nearest stellar black hole is V616 Mon. It's roughly 10 times the Sun's mass and 3,000 light years away. The next closest is the prototypical Cygnus X-1, 15 times the mass of the Sun and at a distance of 6,100 light years. However, we won't have the technology to visit a black hole, even using miniaturized space probes, for many decades, so any discussion about humans falling in is hypothetical. The nearest massive black hole is 4 million times the mass of the Sun, at the center of our Milky Way galaxy, 27,000 light years away. The nearest supermassive black hole is at the center of the giant elliptical galaxy M87, 60 million light years away in the Virgo Cluster. This monster tips the scales at a hefty 5 billion times the mass of the Sun.

However, massive black holes aren't as extreme as you might think. The formula for the Schwarzschild radius, which defines the event horizon, is $R_S = GM/c^2$, so the size of the event horizon is proportional to mass. It's 300 million kilometers, or twice the Earth–Sun distance, for a quasar black hole 100 million times more massive than the Sun. Size increasing linearly with mass means that density within the event horizon decreases according to the square of the mass. The stellar black hole 3 times the mass of the Sun has a density 10,000 trillion times the density of water, while the black hole at the galactic center has a density only 1,000 times higher than water. A quasar black hole 100 million times the Sun's mass has a density only 10% of that of water, and the largest black holes have densities 10,000 times lower still. How scary is a black hole when it's less dense than the air we're breathing!

Let's think about that for a minute. If you took space the size of the Solar System and filled it with air, it would be a black hole. And if you could make an ocean big enough, that black hole would be buoyant and rise like a bubble.

Crossing the event horizon of a massive black hole would likely be much less dangerous than entering a stellar black hole. For one thing, spaghettification would be far less likely. The acceleration due to stretching force declines rapidly as the mass of the compact object increases. At the event horizon of a 100-million-solar-mass black hole, that acceleration would be orders of magnitude less than Earth's gravitational acceleration. An intrepid voyager would cross the event horizon without feeling a thing.

This suggests the ultimate adventure for a space traveler of the far future. Find yourself a black hole; anything bigger than 1,000 times the mass of the Sun will do. Gather your friends and family and position them in a spaceship at a safe distance. They will think of this as your final farewell since nobody can escape a black hole. Then put your spaceship on a free-fall course toward the event horizon. As you approach the event horizon, give a casual wave. Your friends will see your image stretched and distorted. It will also redden as photons struggle to escape the intense gravity of the black hole. You'll see and feel nothing unusual as you pass through the event horizon to an intriguing but unknown fate. Friends and family will be treated to a final tableau of you with your hand poised in mid-wave, the image fading to red and frozen for eternity.

• • •

Let's review the road we have traveled.

Although some early scientists dreamed of black holes, it took a bold new theory of gravity to predict them. Their properties are so bizarre that even the architect of the theory, Albert Einstein, didn't believe such monsters existed. Physicists were energized by the idea of black holes and they redoubled their efforts to reconcile the theories of gravity and the quantum world.

At that point, it was up to the observers. Not everything we can dream and scheme and calculate is real. Black holes form whenever a massive star dies, but they're invisible to the eye, so can only be seen when they orbit a visible star. After several decades of painstaking work, several dozen binaries were found, in which the dark member of the system was so massive that it had to be a black hole. The observations were convincing. Theorists who had bet against the existence of black holes paid up.

Meanwhile, astronomers were accumulating evidence that galaxies are more than just large assemblages of stars. The centers of some galaxies contain swirling hot gas and sources of intense radio and X-ray emission that can outshine the entire galaxy and be seen across most of the universe. The radiation is powered by gravity from black holes millions or even billions of times the mass of the Sun. It's an irony of astrophysics that something so dark can lead to so much light. Our own galaxy harbors a massive black hole, dark because it's slumbering between meals, diagnosed by a swarm of stars that orbit it at millions of miles per hour.

Theorists predicted that all galaxies should harbor massive black holes. With the help of tools like the Hubble Space Telescope, astronomers confirmed the prediction, locating black holes that were inactive and dark and others that were voraciously consuming gas and shining brightly. They've weighed black holes by the thousand. This research has taken away black holes' shock and awe, and given them an air of inevitability—which doesn't make them any less amazing.

Now it's time to explore the implications of black holes. We'll look at their life story and their role in the evolving universe going back to the big bang. We'll learn how they can be simulated in a computer and ask whether or not they could ever be created in a lab. We'll see how they can be used to test our theory of gravity and how we've detected the ripples in space-time that result when they merge. Finally, we'll look at the fate of black holes over near-infinite spans of cosmic time.

PART B

Black Holes, Past, Present, and Future

What is the life story of black holes? Astronomers speculate that some black holes may have been created soon after the big bang, when the infant universe was hot and dense. Since then, small black holes have formed as massive stars die, and large ones have grown by guzzling gas at the centers of galaxies and by combining when galaxies merge. In the second part of this book we look at how black holes of different sizes are formed and grow. Now that their existence is no longer in doubt, astronomers are designing observations to probe closer and closer to the event horizon. Researchers have also learned how to explore the properties of black holes within the safety of a computer simulation.

Black holes are the ultimate proving grounds for the theory of gravity. They allow us to put general relativity to the test as never before. Most of the excitement in black hole research in the next decade will come from the detection of gravitational waves, ripples in space-time that are a central prediction of general relativity. When black hole mergers were first detected a few years ago, a new field of astrophysics was launched. Gravity wave detectors will soon be able to detect black hole mergers across the observable universe, with an event every week. If humans survive that long, our distant descendants will be treated to a close-up view of the merger of the black hole in our galactic center with a similar black hole in the Andromeda galaxy.

We close by looking at how black holes grow and are eventually starved as the universe expands and galaxies dissipate. Even the largest black holes will one day evaporate in a whisper of Hawking radiation. Nothing lasts forever. Not the universe, and not black holes.

5.

THE LIVES OF BLACK HOLES

THE UNIVERSE CONTAINS black holes ranging from city-sized objects with the mass of a star to Solar System–sized objects with the mass of a galaxy. How are black holes born and how do they live their lives? The story starts with the big bang and continues with violent star death and mass convergence at the centers of galaxies. With a mixture of observation, theory, computer simulations, and a dose of speculation, astronomers have pieced together the history of black holes. They have even considered the question of whether the universe itself is a black hole.

Seeds of the Universe

The early universe was chaotic and unstructured. Although the universe has been getting lumpier as gravity forms planets, stars, and galaxies, it was never *perfectly* smooth. Just after the big bang, there were slight nonuniformities, and since the average density of the universe was extremely high, gravity in these regions would have been very strong. The seeds for the formation of galaxies thus date back to the early universe. But that's not all. In the same year he predicted the radiation that bears his name, Stephen Hawking and his student Bernard Carr wrote a paper about black holes that may have formed in the very early universe: primordial black holes.[1] They argued that even if the density variations that occurred just after the big bang were small on average, variations in some regions could have been large enough to create a gravitational attraction that exceeded the force of cosmic expansion. In those locations, gravitational collapse would occur and a black hole could form. This process can make black holes of almost any mass. Could Hawking's primordial black holes have been the seeds of the universe?

The earliest black holes would have formed at "Planck time," the period 10^{-43} seconds after the big bang, when the universe was 10^{-35} meters across.[2] Black holes that formed then would have had a mass of 10^{-8} kilograms, about that of a mote of dust. These early black holes couldn't grow due to the rapid expansion of the universe, and thus quickly evaporated. Any black hole that formed less than 10^{-23} seconds after the big bang with a mass less than 10^{12} kilograms would have evaporated by now, but later-forming and more massive ones might survive to the present day. A primordial black hole that formed a second after the big bang would have had a mass of at least 100,000 solar masses, not far short of the Milky Way's central massive black hole.

Another intriguing theory holds that primordial black holes might persist in an unexpected form. For the last forty years, astronomers have been dealing with the problem of dark matter. Stars in galaxies of all kinds move too quickly to be explained by the gravity of the stars themselves. There seems to be an extra component of mass, 5 or 6 times more than the sum of all the stars, that's holding galaxies together.[3] This dark matter exerts gravity, but it doesn't emit light or interact with radiation in any way. Gravitational lensing data shows that dark matter also fills the space between galaxies. What if dark matter was made of primordial black holes? It's an attractive possibility. In theory, primordial black holes, like dark matter, should be found throughout the cosmos, and positing them as the source of dark matter avoids invoking a new fundamental particle that's outside standard physics (and hasn't yet been seen by accelerators).

Unfortunately, careful observations have eliminated most of the ways primordial black holes might exist, including as dark matter. When a black hole evaporates, it releases a torrent of gamma rays; by the 1980s, NASA had gamma ray detection satellites in orbit, but they didn't see the expected signature. Gravitational lensing rules out widely distributed black holes from galaxy mass down to Earth mass. Recent theoretical work has closed the door on the last mass window, from 10^{14} to 10^{21} kilograms, or from the total carbon mass in the atmosphere of the Earth up to the mass of a small Solar System moon.[4] Primordial black holes can't be abundant enough to account for dark matter, but that doesn't mean they don't exist in some form. They're predicted by cosmological theory and have the potential to illuminate our understanding of the early universe. The search continues.

First Light and First Darkness

Just a few seconds after the big bang, conditions no longer favored the formation of primordial black holes. The universe at that moment was an almost perfectly smooth cauldron of high-energy particles and photons, with density variations of less than 0.001% from one place to another. A few minutes after the big bang, the temperature had dropped to the point where atomic nuclei could form. Fusion converted a quarter of the mass of the universe from hydrogen to helium, with trace amounts of lithium and isotopes of hydrogen and helium. This took no longer than the time needed to boil an egg. The temperature was 10 million degrees; you'd need X-ray vision to see the universe then.[5]

The universe continued to expand and cool. The next important milestone occurred at about 50,000 years, when the energy densities of matter and radiation were equal. Thereafter, the energy density of radiation fell more rapidly than matter density as photons were red-shifted by expansion. As a result, gravity exerted its grip and the tiny density variations could start to grow. The temperature of the universe was 10,000 degrees. If there had been anyone there to see it, it would have been glowing blue. About 400,000 years after the big bang, the temperature had fallen to 3,000 degrees and electrons joined nuclei to form stable atoms. Radiation traveled freely for the first time, the "red fog" lifted, and nascent structures came into view.

It was still early days. Compared to an age of 13.8 billion years, 400,000 years is the blink of an eye—the equivalent of the first ten hours of a forty-year-old human's life. As it expanded, the universe faded from view; its radiation slid from dull red to invisible infrared. This was the beginning of the Dark Ages.[6] The Dark Ages continued until the first stars and galaxies formed, 100 million years or so after the big bang, so this entire era lies within the first 1% of the age of the universe.

Intriguingly, although the first fraction of the lifetime of the universe was dark, it might not have been lifeless. From 10 to 20 million years after the big bang, the universe had a temperature between the boiling point and the freezing point of water. The present-day universe is extremely cold, and biology as we know it can only exist in slender habitable zones near stars, or perhaps in colder locations under the surface of a planet or moon where water is kept liquid by pressure from above and radioactive heating from below. But there was a time when the entire universe had a habitable temperature. What isn't clear is whether or not the rare early stars could have made enough carbon for biology to develop and enough heavy elements to form a planet for that biology to inhabit.[7] It's also doubtful if 20 million years is enough time for life to evolve from simple chemical ingredients.

Some of the most important questions in cosmology center on the Dark Ages. When did it end? Which formed first, stars or galaxies? How were the formation processes affected by the absence of heavy elements? What's the best way to detect first light in the universe? And most important for our narrative: what kind of black holes were first to form?

Let's assume for a moment that dark matter is a new type of fundamental particle predicted by theories that unify three of the forces of nature. As an ingredient in cosmology, dark matter is fairly simple: it exerts gravity but doesn't interact with light or any other form of radiation.[8] There's 6 times more dark matter than normal matter, so it dictates the formation of structure in the universe. As dark matter is concentrated by gravity, small or low-mass clumps begin to emerge. The first structures to form, 100 million years after the big bang, as the Dark Ages end, have 10^6 solar masses of dark matter. That's the mass of a tiny dwarf galaxy in the present-day universe. As time passes, these clumps merge to form bigger and bigger clumps. Each clump of dark matter contains a "puddle" of normal matter one-sixth the mass of the dark matter, and that gas collapses into the center of the dark matter gravity "pit." When it collapses, stars form and first light is triggered. In this "bottom up" scenario, small objects form before large objects; stars form before galaxies (Figure 37).[9]

The universe was very different as the Dark Ages ended and the first lights flickered on. It was 30 times smaller, 30 times hotter, and 30,000 times denser than it is now. The other major difference was the lack of elements heavier than hydrogen or helium. The process of star formation depends on heat being radiated away so that gravity can make the gas cloud collapse. Carbon and oxygen have spectral transitions that very efficiently carry away energy. The lack of these elements in the early universe means that star-forming clouds were hotter and more massive. In the local and current universe, the upper limit of star mass is about 100 times the mass of the Sun. In the early universe, the first stars probably ranged up to 200–300 times the mass of the Sun. Long ago, million-solar-mass clumps of dark matter formed stars that on average were dozens of times more massive than the stars forming now in the neighborhood of the Sun.

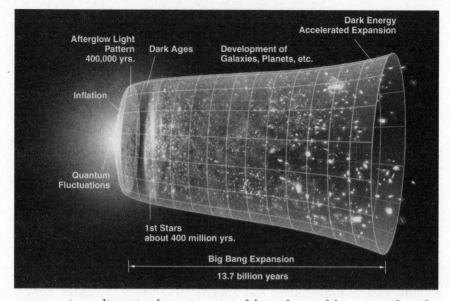

FIGURE 37. A two-dimensional representation of the evolution of the universe from the big bang 13.7 billion years ago. The exponential early expansion gives way to slower expansion. The universe is dark from 400,000 years to a few hundred million years after the big bang, when the first stars and black holes form. Massive black holes grow due to galaxies merging and gas falling into galaxies. In the last 5 billion years expansion has accelerated due to dark energy. *NASA/WMAP Science Team*

The lives of the first stars were short. They raced through their nuclear fuel in a few million years. In computer simulations the most massive stars explode as supernovae, leaving nothing behind, or they collapse directly into black holes 20 to 100 times the mass of the Sun. Like the first stars themselves, the black holes they leave behind are more massive than the black holes we find nearby in the Milky Way.

Everything you've just read is based on theory and computer simulations. What about the observational search for first light? There are two approaches, and they're both like searching for a needle in a haystack, since the first stars are scarce and the universe has been forming stars steadily for 14 billion years. One approach is to look for stars in the Milky Way that are made of only hydrogen and helium, which would mean they formed from gas that had not been "polluted" by any previous generation of stars. In 2012, a group at the European Southern Observatory followed up a faint star from the Sloan Digital Sky Survey and found that it had a heavy element abundance 200,000 times less than the Sun.[10] At 13 billion years old, it's the best candidate for a primordial star.[11]

The other approach is to find stars without heavy elements in a distant galaxy. In 2015, another European group saw pristine stars in a galaxy at a redshift of z = 6.6, meaning the light dates from less than a billion years after the big bang. David Sobral of the University of Lisbon, the lead author, named the galaxy CR7, for Cosmos Redshift 7, but also for his favorite soccer player, Cristiano Ronaldo. Sobral said, "It doesn't really get any more exciting than this. This is the first direct evidence of the stars that ultimately allowed us all to be here by fabricating heavy elements and changing the composition of the universe."

Black Hole Birth by Stellar Cataclysm

In July 1967, two American Vela satellites detected gamma ray pulses. A Cold War invention, these satellites had been launched to detect Soviet violations of the 1963 Nuclear Test Ban Treaty.[12] The public didn't know at the time, but the U.S. government was in a heightened state of preparedness for war.

Luckily, a team from Los Alamos National Laboratory showed that the gamma ray flashes were unlike the signature of a nuclear weapon, and deduced that the sources were located far beyond the Solar System. In 1973, this discovery was declassified and published as a research paper.[13] Yet the mystery deepened. There was one gamma ray burst every day somewhere in the sky. For a few seconds, these sources outshone the entire rest of the universe in gamma rays. But they also faded quickly, lasting from a few milliseconds to about thirty seconds. The positions measured by gamma ray satellites were too crude to allow follow-up of the bursts, and the distribution was random so it gave no clue to their origin.

The breakthrough came in the late 1990s, when a rapid response X-ray satellite started taking data in orbit. It could quickly pivot to catch lower-energy X-rays from the gamma ray event, and accurate X-ray positions allowed optical astronomers to catch the fading afterglow. Spectroscopy showed that the objects responsible for the bursts were in distant galaxies, billions of light years from the Earth. The large distances meant that the bursts must be phenomenally bright. One event in 2008 would have been visible to the naked eye for thirty seconds, despite occurring halfway across the universe. The light that appeared briefly in 2008 was emitted 3 billion years before the Earth formed.[14] Another burst that was seen in 2009 was in a galaxy at a redshift $z = 8.2$, so that event occurred when the universe was just 4% of its present age.[15] The strongest bursts put out 1,000 times more energy than a supernova, up to 10^{44} joules. That's the lifetime energy output of the Sun, emitted in a second rather than spread over 10 billion years!

When a gamma ray burst goes off, catching the optical afterglow is the only way to measure redshift and luminosity, which can help tell us how old an object is and point us toward how massive it might be. A few years ago, I was on the 6.5-meter Multiple Mirror Telescope on Mount Hopkins in Arizona when I got an Internet alert. NASA's Swift satellite had detected a gamma ray burst and the call went around the world to take a spectrum. It was 3 a.m. but I set aside my coffee; nothing wakes you up like chasing a stellar cataclysm. Within minutes we were at the position. Nothing was visible on the TV monitor, so we worked blind and hoped for a signal. The next day the reduced data showed a ragged trace with hints of emission lines, but they weren't strong enough to measure a redshift. The following night it had faded beyond detection. In astronomy sometimes you have to settle for the thrill of the chase.[16]

Astronomers believe gamma ray bursts are the calling card of a newly formed black hole.[17] The thousands of events studied so far divide into two populations: high-luminosity long-duration events, and low-luminosity short-duration events. The most luminous bursts are due to the collapse of the rotating core of a massive star, typically more than 30 times the Sun's mass, which forms a black hole. Matter near the star core rains down onto the black hole and swirls into an accretion disk. The falling gas creates twin jets along the rotation axis, which travel at 99.99% of the speed of light and pound their way through the star's surface to radiate as gamma rays. Much of the gravitational energy is released in the form of neutrinos rather than photons (Figure 38). The quicker bursts are thought to be caused by the merger of two neutron stars, or the merger of a neutron star and a black hole. Either of these cases results in a single black hole. Most of the energy of the merger is released in the form of gravitation radiation, space-time ripples that radiate outward at the speed of light as predicted by general relativity. Matter falling into the newly formed black hole forms an accretion disk and releases a burst of energy.

A hypernova is an even more extreme type of black hole formation event. It releases hundreds to thousands times more energy than the normal death of a massive star in a supernova. The record holder is an

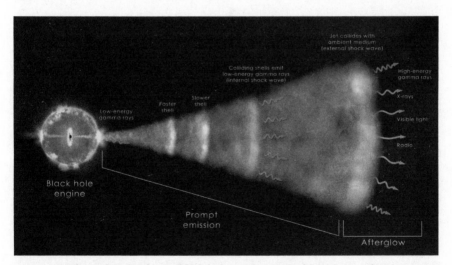

FIGURE 38. The violent birth of a black hole can be marked by a burst of gamma rays. As energy emerges in relativistic jets along the polar axis of the black hole spin, a blast wave creates extremely high-energy electromagnetic radiation. For a few seconds, such explosions are the brightest objects in the universe in gamma rays, detected at distances of billions of light years. *NASA/Swift Science Team*

explosion reported in 2016 that was half a trillion times brighter than the Sun.[18] Imagine 20 times more light than all the stars in the Milky Way, crammed into a space 10 miles across. This explosion was the biggest recorded since the big bang, a mind-boggling fact that challenges any physical theory of the energy it released.

Such an enormous explosion raises an uncomfortable question: is the Earth at risk from a stellar cataclysm? In other words, though we don't have to worry about falling into a black hole, should we worry about a black hole reaching out to smack us? The good news is that these events are rare, about one per galaxy per million years. Also, the radiation is concentrated in twin beams, and the explosions are randomly oriented in space, so 99.5% of them will miss us. That takes the average incidence down to one per galaxy per 200 million years. The bad news is that if we happen to be in the line of fire and the blast is within a few thousand light years, the Earth and its biosphere will get hammered by high-energy radiation. Gamma rays would deplete the ozone layer by 75%, spiking mutation rates. The total effect on the ecosystem is difficult to estimate, but one group argues that the late Ordovician mass extinction 450 million years ago was caused by a gamma ray burst.[19] The evidence from the extinction is consistent with ozone depletion and a loss of surface species, but there's no way astronomers can pinpoint an explosion that ancient because the black hole is all that's left.

There's more impressive evidence that a milder event occurred within recorded history. In 774 AD, the Western world was a patchwork quilt of small warring states. Charlemagne was consolidating his kingdom with conquests of Tuscany and Corsica. In Japan, where Buddhism was rapidly becoming the state religion, Empress Koken had a million prayer scrolls made; they're among the oldest printed works in the world. Carbon dating shows that the trees that were felled to create these scrolls had experienced a sharp increase in their carbon-14 to carbon-12 ratio.

This spike is Exhibit A in making the case that the Earth was irradiated by a gamma ray burst about 1250 years ago. Carbon-14 is radioactive and decays into nitrogen. The fact that it exists at all is due to cosmic rays, high-energy particles from space, hitting nitrogen in the atmosphere. This process maintains a constant low level of carbon-14, but a spike by a factor of 10, as observed in the scrolls, must have had an additional external cause. Exhibit B is a rise in carbon-14 in trees in Europe and America, although the date is harder to pin down. Exhibit C is a small jump in radioactive beryllium-10 around that time.[20] Beryllium-10 is created when high-energy particles hit an exposed surface; its concentration is used to date glacial advances, lava flows, and other geological events in rocks as old as 30 million years. None of this can be explained by a solar flare. Nor can it be explained by a supernova, because any supernova that close would have been visible in the daytime sky and none was recorded in medieval manuscripts. That leaves a gamma ray burst. At a distance of about 5,000 light years, it would have dumped 200 megatons of gamma ray energy into the Earth's atmosphere. The afterglow only lasted a few days so, even though it would have been visible to the naked eye, it's likely that nobody noticed it or thought to record it.

Meanwhile, astronomers have their eye on a massive star called WR 104, which is 8,000 light years away and likely to die with a violent core collapse sometime in the next few hundred thousand years. We cannot measure its orientation in space, so when it goes off, we'll have to hope that one of its powerful jets isn't pointing in our direction. Astronomical timekeeping is crude enough that this timescale isn't completely reassuring. It may go off much sooner. In the meantime, there are better things to lose sleep over.

Finding the Missing Links

We've talked about two different kinds of black hole. One forms from the death of a massive star, when a star that started its life at between 8 and 100 times the mass of the Sun leaves behind a dark object between 3 and 50 times the mass of the Sun. The other forms at the center of a galaxy, and ranges from a few million times the mass of the Sun in non-active spiral galaxies such as the Milky Way up to a few billion times the mass of the Sun in giant elliptical galaxies such as M87. That leaves a huge gap in mass: five powers of ten, from a few dozen solar masses up to a few million. Do intermediate-mass black holes exist?

A small set of objects have been found that push into that gap from the low end. Recall that Arthur Eddington worked out the limit of the brightness of a black hole. The faster black holes eat, the brighter they shine. But even if a black hole is feasting on plentiful gas from a companion in a binary system, there's a limit to its brightness. The pressure of radiation blazing off the accretion disk counteracts the black hole's gravitational pull, so at some point excess gas trying to fall in is blasted back into space. That's called the Eddington limit. Thirty years ago, a rare class of ultra-luminous X-ray sources (ULXs) was discovered. They pump out a million times more X-ray energy than the Sun's total power and they're bright enough to be seen in galaxies millions of light years away. According to the Eddington limit, these black holes must be hundreds or thousands of times more massive than the Sun—slap bang in the middle of the mass gap.[21]

Luminous X-ray binaries are important for another reason. Some are scaled-down versions of quasars. The exotic binary system SS 433 is 18,000 light years away, in the constellation of Aquila. A bloated blue star orbits a black hole every thirteen days and siphons gas onto the accretion disk around the black hole. Some hot gas falls into the black hole while the rest is funneled into twin jets that shoot out along the rotation axis of the black hole. The gas moves at a quarter of the speed of light, covering a mile in 20 microseconds.[22] SS 433 is the archetypal microquasar (Figure 39). Microquasars have all the ingredients of a quasar—spinning black hole, accretion disk, intense high-energy radiation, relativistic jets—but are scaled down by a factor of a million.

There are only 100 known microquasars in the Milky Way, but they're very helpful in modeling and understanding the extreme astrophysics of quasars.[23] The fueling timescale for a quasar is far longer than a human lifetime while for a microquasar it's a few hours, so it's easily observable.

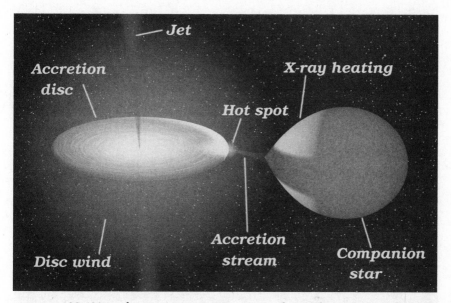

FIGURE 39. SS 433 is a binary star system consisting of an early-type star and a black hole. This is a visualization, since the system, at a distance of 18,000 light years, cannot be seen in detail. Gas from the companion has a lot of angular momentum so it flows on the black hole via an accretion disk. Doppler shifts in both optical and X-ray spectral lines show the speed of the relativistic jets. This kind of object is a miniature version of the astrophysics taking place in the center of a quasar. *M. Rupen, R Hynes/NRAO*

What about pushing down into the gap from the high end? Let's return to a central insight of the past few decades: every galaxy has a heart of darkness. Quasars and active galaxies are rare. The central black holes in most galaxies are inactive most of the time, so they can only be detected by their influence on the stars near the center of the galaxy. As astronomers gathered more data on black holes in nearby galaxies, they saw a striking correlation. The mass of an inactive central black hole is accurately predicted by the velocity spread—the range of motions that indicates total mass—of the old stars in a galaxy.[24] This correlation is puzzling. These black holes only affect the very central regions of a galaxy, and stars in the galaxy outweigh them by a factor of 500. Why are these disparate quantities related?

Astronomers aren't sure, but the correlation has recently been extended down to dwarf galaxies and even globular clusters with black holes of a few thousand solar masses (Figure 40). Elliptical galaxies are large and made up almost entirely of old stars, so they have the most massive black holes. Spirals like the Milky Way have fewer old stars, mostly gathered in small central bulges, so they harbor more modest black holes.

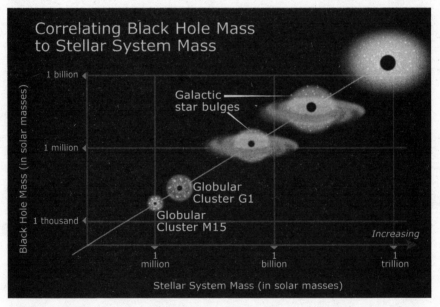

FIGURE 40. There is a fairly tight correlation between the mass of old stars in a galaxy and the mass of its central black hole, extending over a factor of 100,000 from dwarf galaxies to giant elliptical galaxies. Globular clusters within the Milky Way extend this correlation down to a few thousand solar masses. This correlation shows that the central black hole only contains a few tenths of a percent of the stellar mass of a galaxy. *A. Field/NASA/ESA*

Observation of smaller black holes is challenging, and pushes telescopes and detectors to their limits. The best targets are globular clusters, the spherical clouds of stars that orbit the haloes of large galaxies. With a few hundred thousand to a few million stars, the correlation described above predicts black holes of several thousand solar masses. Detections have been claimed, but none has survived skeptical scrutiny. Nonetheless, a handful of objects have filled in the gap. In 2012, for example, a 20,000-solar-mass black hole was seen in the dwarf galaxy ESO 243-29, and in 2015, a 50,000-solar-mass black hole was found in the dwarf galaxy RGG 118.

The most significant discovery of an intermediate black hole came late in 2015, when Japanese radio astronomers spotted a swirling gas cloud only 200 light years from the center of the Milky Way. They tracked the rotation with spectral lines from eighteen different molecules and deduced the presence of a dark object 100,000 times the mass of the Sun. The discovery supports the idea that black holes grow the same way aggressive companies do, by mergers and acquisitions.[25] Millions of years from now, when the 4-million-solar-mass black hole at the center of our galaxy consumes this intermediate creature, the central beast will grow by 2.5% and we can imagine it emitting a burp of satisfaction. This burp will be registered as a pulse of high-energy radiation hitting the Earth 27,000 years later.

Simulating Extreme Gravity in a Computer

Einstein thought about gravity in an entirely new way. Gravity was not, as Newton posited, pulling or tugging objects around in space. An object moving in response to gravity was following the shortest path, called a geodesic, through curved space-time. An astronaut slowly falling toward a spacecraft is simply following the curvature of space-time. The Moon circles the Earth because the shortest path through space-time brings it back to the same point in space. A two-dimensional version of this happens every time you take a long plane flight. Imagine flying from Los

Angeles to Madrid. Even though these cities are at the same latitude, the plane doesn't fly due East. It heads north and flies over the southern tip of Greenland before heading south. It follows the shortest distance between those two points, as you could confirm if you stretched a string over the surface of a globe. The pilot doesn't need to turn left or right; the route is a straight line on a curved two-dimensional surface.

The simplest form of general relativity is written as $G = 8\pi T$, where G is the space-time curvature at a point and T is the mass at that point (technically, it's mass-energy, but since energy only has a tiny amount of equivalent mass according to $E = mc^2$, for astronomy situations it works to just consider the mass). This little equation applies at all points in space, and encapsulates everything we need to know about gravity.[26]

However, this elegant equation is the highly compact form. It's useless for solving any real problem. To apply general relativity to something like a black hole, the full expressions have to be used, which expand into ten different equations, each with many terms. To solve these equations involves a fiendish amount of difficult algebra and calculus. And to understand what happens when two black holes of different masses merge, every term in every one of Einstein's equations has to be used. Written out, it's 100 pages of dense math. No simplification is possible.

In the 1990s, after rapid mathematical and computational advances, numerical relativity took off. Approximations to Einstein's equations were developed. They centered on methods to separate space and time, sampling the space so finely that Euclidean geometry could apply. Computation uses an "adaptive mesh," where the space grid is coarse where gravity is weak and flat, and fine where gravity is strong and curved. The grid is adjusted continuously as the situation evolves. Computer speed is measured in floating point operations per second, or flops. An IBM 7090 computer, which was state-of-the-art in 1962, had a speed of 100,000 flops. In 1993, the fastest computer was a million times faster. Now it's a million times faster still, or a staggering 10^{18} flops.[27] The National Science Foundation spurred this research with "Grand Challenge" grants to simulate a binary black hole collision.[28] The numerical work turned up some surprises. Such a merger can produce an enormous amount of gravitational radiation: 8% of the total mass of the black holes. Also, when two black holes merge the resulting black hole can get a "kick" speed of 400,000 mph, enough to eject it from any galaxy.[29]

Let's use a canvas as a visual metaphor for something that can't be seen: space-time. The canvas is painted with gravity. So far we have just stretched the canvas of empty space-time (Figure 41). General relativity is a geometric theory of gravity, so if there's mass anywhere, the space-time canvas is curved, and it can contain punctures and tears and folds. The canvas is three-dimensional, so impossible to visualize. But the canvas isn't the whole story. A black hole in the real universe is surrounded by radiation and hot gas and high-energy particles and magnetic fields.

We're talking about three levels of difficulty. Level 1 is complex interactions between particles and radiation. Level 2 adds in magnetic fields. Level 3 includes gravity. At this point, researchers are exploring a technique called general relativistic magnetohydrodynamics, which is a real conversation stopper at cocktail parties. To use a game analogy, the three levels are as checkers is to chess is to Go. In this spectrum of technical ability, I'm nifty at checkers, I can hold my own at chess, but I'm completely flummoxed by Go. The full numerical treatment aims to represent the complex astrophysics of not just the black hole, but also the accretion disk and the twin jets.[30] This is the state of the art in black hole simulation. Fewer than 100 people in the world have the technical skills to do this work.

Computers can simulate small black holes, but what about the large ones that live at the centers of galaxies? For this we meet Simon David Manton White, Fellow of the Royal Society and director of the Max Planck Institute for Astrophysics in Garching. He is a magician with gravity in a computer, so we'll dub him the Sorcerer. The Sorcerer has sad eyes, a tidy mustache, and a graying mop of curly hair. He looks weary, but you'd be weary too if you'd made the universe from scratch.

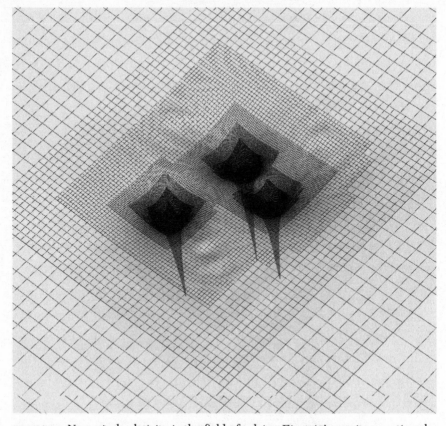

FIGURE 41. Numerical relativity is the field of solving Einstein's gravity equations by computational methods, in a complex and realistic astrophysical setting where exact solutions to the equations are impossible. One requirement is to work at different size scales. For a binary black hole this ranges from the orbital scale to the event horizon scale. As illustrated here, one powerful method is called adaptive mesh, where the sampling of space-time in the calculations automatically adjusts to the strength of the local gravity. *GRChombo collaboration/Lawrence Berkeley National Laboratory*

The Sorcerer studied for his PhD in Cambridge with Donald Lynden-Bell, black hole pioneer and visionary. He has over 400 refereed publications and more than 100,000 citations, eye-blinking numbers that place him in the rarified stratosphere of his subject. He's a world expert on the properties of dark matter and on the formation of structure in the universe.[31]

Here's how you make the universe in a computer. Set up a three-dimensional grid of space. Add normal matter and dark matter in the correct proportions. Switch on gravity. Make the space expand according to the big bang model, and watch a filigree of large-scale structures congeal from the initially smooth distribution of mass. A large number of "particles" stand in for astronomical objects. For example, a million particles might be used to represent a star cluster with one particle per star, but no simulation has enough particles to represent a galaxy with one particle per star, or the universe with one particle per galaxy, so in practice a particle represents a varying amount of mass.[32] As an analogy, imagine using a million particles to model human populations. In a model of the world, each particle would represent 7,500 people, or the number of people living in a village or a small rural region. No finer detail is possible. But the same model could represent the gory details, with one particle per person, for a small state in the U.S. like Rhode Island or a moderate-sized city like Austin, Texas.

The computational demands rise rapidly as the number of particles increases, so White and other programming gurus use tricks to speed up a simulation dramatically.[33] After all, nobody wants to wait 13.8 billion years for the outcome. White's simulation was known as the Millennium Run because it was the first powerful mockup of a large chunk of the universe after the year 2000.

These simulations only include gravity. But galaxies have gas as well as stars, and gas behaves differently from stars. When two spiral galaxies collide, the stars and the dark matter particles almost never collide, so those components of the galaxies pass through each other. But the gas components crash into one another, heat up, glow brightly, and form stars. Gas behaves more like a fluid than a set of particles. To deal with the gas, the simulators mimic gas behavior with smoothed-out particles that have a probability distribution instead of a single location.[34] They also layer in physics, via equations to include important but small-scale details like supernova explosions and black hole formation. Listen to Simon White talk about his landmark cosmological simulation:

What was novel in the original Millennium Run was first of all the overall size, which was roughly a factor of 10 larger than previous calculations. And also the fact that we implemented techniques which allowed us to follow the actual formation of visible galaxies in a rough but physically based manner. We were able to predict not only the distribution of the unseen dark matter component of the universe, but also where the things we can actually see should be and what their properties should be. . . . There already have been surprises. One was the realization that to understand the properties of the visible galaxies we must understand the effects of the black holes in their centers. The actual population of galaxies was shaped by the development of the black holes in their cores. It is not true that this small object in the center is divorced from the rest of the galaxy, even though the black holes only contain a tenth of a percent of the stellar mass of the galaxy, a very tiny fraction.[35]

The Millennium Run was completed in 2005.[36] It used 10 billion particles to simulate a cube of the universe 2 billion light years on a side. (The results required 25 terabytes of storage.) That's not the whole universe, but it's large enough to be a "fair sample" that encompasses the largest structures gravity can form in 14 billion years. Hundreds of scientific papers have been published based on the simulation. The current state of the art is the Illustris Simulation.[37] Following Moore's law, the gain in computing power that results from transistors getting smaller, the size of the best simulation doubles every twenty months. By the end of 2017, the simulators had cracked the trillion-particle barrier. With Illustris, for the first time it's possible to realistically model a substantial fraction of the universe down to a level of detail that can resolve the structure of an individual galaxy. A computer can now track the fueling and growth of millions of supermassive black holes across 13 billion years.

Like many theorists in astronomy, Simon White's initial training was in mathematics. He recalls weighing his choices for graduate school: "I had two options in Cambridge. One was to do theoretical fluid mechanics, aerodynamics and this kind of thing. There the students were in the building in the center of Cambridge, in basement offices with no windows. The other option was astrophysics. The astrophysics center was outside the town, and it was a building with a lot of windows. There also were trees and cows across the road. I thought astrophysics looked a little better."[38]

How Black Holes and Galaxies Grow

The lives of black holes and galaxies are intertwined. A supermassive black hole occupies a tiny fraction of the volume of a galaxy and it has a tiny fraction of the galaxy's mass. Yet we've seen that every galaxy has a black hole and the mass of the black hole is tightly coupled to the mass of the stars in the entire galaxy. What does that imply about how black holes and galaxies grow together over cosmic time?

The glory days of quasars are long gone. We can detect supermassive black holes lurking in nearby galaxies, but they are mostly quiet, like the black hole in our own galaxy. One in 100 is mildly active, and one in a million is a quasar. Surveys at optical and X-ray wavelength can track the brightness of quasars going back in time. The peak of quasars occurred at redshifts of $z = 2$ to 3, about 11 billion years ago, or 2 to 3 billion years after the big bang. They were thousands of times more active than they are now. The ancient night sky was quite different from our night sky. The universe was 4 times smaller and galaxies were merging and forming stars rapidly. Hundreds of galaxies would have been visible to the naked eye, as opposed to the three we can see now,[39] and the nearest quasar would have been 100 times closer, also visible to the naked eye. Observations tell the story of a rapid rise in quasar activity followed by a slow decline.

All large galaxies have supermassive black holes, but it doesn't follow that they all show quasar activity. How do we know that quasar activity is episodic rather than a property of a particular set of galaxies? Answering this question is hard, because astronomers can't stare at particular galaxies to see how they evolve. They do surveys that capture a large number of galaxies at all epochs, and use that data to take a snapshot of activity at a particular epoch.

This has been my research area for the past decade. The goal is to understand how black holes and galaxies grow and intersect with one another. I enjoy astronomy research because it hasn't gone the way of physics, with collaborations a thousand strong and instruments that take a decade to build. It's still possible to go to the telescope for a few nights with a graduate student and a good idea and make an impact.

That was how Jonathan Trump and I came to be in the foothills of the Andes, watching the sky darken over the cordillera and preparing our observing list. We were aiming for the sweet spot of black hole growth, the time between 3 and 10 billion years after the big bang when galaxies did most of their merging and black holes did most of their feeding. In particular, we were trying to identify the lower boundary to nuclear activity. How feeble could the accretion rate be for the black hole to light up as a quasar? We managed to find black holes 10 billion light years away that were as quiescent as the black hole in the center of our galaxy. At our disposal was technology that obliterated the photographic plates I'd used in Australia thirty years earlier. Not only could we discover 300 quasars in a good night, but we could measure their black hole masses as well.

Jon was the energetic apprentice while I was the grizzled veteran— but in practice, we often switched roles. Time hadn't dulled my enthusiasm for research and occasionally I made blunders in my haste to gather data, while he reined in my excesses and kept a steady hand on the tiller of the telescope. We hit a patch of clouds in the middle of one night, so I tried to be patient. The photons had traveled for billions of years to be captured by our big glass; what difference would a few hours more make? I walked outside and watched the sky clear. To the west were cloud tops all the way out to the Pacific Ocean. To the east the star field was etched by the jagged silhouette of the Andes. A condor wheeled silently overhead.

The last night on the telescope was tinged with exhaustion and sadness. We were treated to a green flash as the Sun set. Astronomers are lucky if they get half a dozen nights on a big telescope in a year. If it's cloudy, they have to come back again the next year. After this run on the telescope, we and it would part company.

We puzzled over scatter plots of our data. Some quasars had hefty black holes but were sputtering feebly. Others had puny black holes but were blazing brightly. The fueling mechanism was enigmatic. We bagged 500 black holes in a week, but that felt like a pinprick in a universe with 100 billion galaxies, each harboring a black hole. The black holes seemed to silently mock us, and they held their secrets close.

We've seen earlier that sustaining quasar-level brightness requires an accretion rate of a few solar masses per year. There are two implications from such a modest fueling rate. First, there isn't much gas in the central regions of most galaxies and black holes rarely swallow stars whole, so that fuel is exhausted in under 100 million years. Gas drizzles onto galaxies from intergalactic space, and gas is added when galaxies merge, but both processes are inefficient now that the universe is large and galaxies are widely separated. Since black hole "fuel" is used up much more quickly than the time taken for the observed quasar population to rise and fall, individual quasars must be "on" a small fraction of the time and "off" a large fraction of the time.

Second, the modest growth rate of black holes means they shouldn't get very big very fast. But they do. The Sloan Digital Sky Survey has been looking for quasars within the first few billion years after the big bang. The oldest quasar it has found so far is a luminous quasar at a redshift of $z = 7.5$. This suggests that supermassive black holes several billion times the mass of the Sun formed and grew during the first billion years after the big bang.[40] That seems to be at odds with a slow, methodical progression from small to large galaxies by mergers. It's also at odds with the maximum rate a black hole can grow as defined by Arthur Eddington a century ago. It's not possible to start with a "seed" mass of 10 solar masses, typical of the black hole left behind when a massive star dies, and get to a billion solar masses in a billion years. To form these ancient, luminous quasars, the seed mass must have been 10,000 solar masses.

Recent simulations have suggested an explanation. In the first wave of galaxy formation a few hundred million years after the big bang, background radiation initially stopped stars from forming. When they did, formation was a rapid, violent process that left behind many small black holes, and in the high-density environment those small black holes merged to form black hole seeds of 10^4 to 10^6 solar masses.[41] Jump-starting black hole growth in this way enables growth to a supermassive level (a billion solar masses or more) in another half billion years.

The concept of feedback helps tie these observations together. A black hole has a symbiotic relationship with its galaxy host. It can't grow or light up like a quasar without a gas supply from the central regions of the galaxy. But when it is active, it puts out so much energy that it drives away gas in the central regions and suppresses star formation. In an active phase of 10 million years, a quasar emits 10^{53} joules of energy. That's roughly the same as the gravitational energy that binds stars in their orbits of a large galaxy. So a quasar clearly has the power to disrupt a galaxy. Feedback means quasars drive out gas and quench their own activity. The gas has to build up again to start a new active phase. Feedback ties together the evolution of the inner region of a galaxy and its central black hole, and so explains the correlation that astronomers see between black hole mass and the mass of stars distributed on much larger scales.[42]

To put it all together, galaxies and black holes entered an intense construction phase during the first few billion years after the big bang. Governed by dark matter, galaxies grew hierarchically, so small things formed first and merged over time to form big things. Star formation and the merger rate peaked and then went into a slow decline as the gas supply ebbed and the universe got larger. The black hole construction project went differently. The deepest gravitational potentials formed the biggest galaxies and the most massive black holes quickly. They're around us now as elliptical galaxies, long ago starved of gas, with dead quasars lurking in their centers. Meanwhile, the shallower gravitational potentials formed midsized galaxies such as the Milky Way, which grew smaller black holes that continued growing and stayed active longer.[43] The halcyon days of galaxy and black hole creation are long over (Figure 42). In the future twilight of the universe, when the last stars are dying and few new stars are replacing them, the only excitement will come from those rare occasions when two mature galaxies collide and their massive black holes merge.

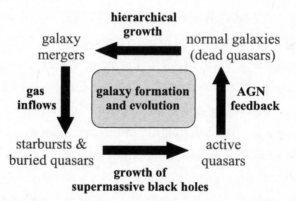

FIGURE 42. There is a complex interplay between black holes and their surrounding galaxies. Galaxies grow hierarchically, going from small to large by mergers, and the central black holes also grow, by a combination of merging and gas falling in. Quasar activity can drive an outflow of gas that quenches the activity, a phenomenon called feedback. Finally, when gas is exhausted or feedback is very strong, the black hole is starved and the galaxy hosts a dead quasar. *P. F. Hopkins/California Institute of Technology*

The Universe as a Black Hole

Is the universe a black hole? There are some superficial similarities. The mass and radius of the observable universe fit the same relationship defined by the mass and Schwarzschild radius of a black hole. And the universe has an event horizon, which is the boundary between galaxies we can see and galaxies we can't see because their light hasn't had time to reach us in the age of the universe.

There are also real differences. At a trivial level, a black hole has an inside—the sealed-off space and time inside the event horizon—and an outside. The universe is defined as all space and time, so it has no "outside." Also, the event horizon of a black hole is a one-way barrier: while no information can escape, we could choose to pass through the event horizon and learn about what's inside. In our accelerating universe, the event horizon, at a distance of 16 billion light years, marks events we will never see, no matter how long we wait. We might see events in galaxies that took place before they crossed the event horizon; subsequent events are forever beyond our view.[44] Ned Wright, an astronomy professor at UCLA, puts it succinctly in his cosmology FAQ: "The Big Bang is really nothing like a black hole. The Big Bang is a singularity extending through all space at a single instant, while a black hole is a singularity extending through all time at a single point."[45] Another way to say this is that our universe had a singularity in the past from which everything arose, while a black hole has a singularity into which things might disappear in the future.

Black holes have also been invoked to explain the existence of the universe. This is speculative cosmology, so fasten your seat belts. The big bang theory depends on an episode of inflation, a period of exponential expansion 10^{-35} seconds after the big bang, during which the universe ballooned from a size smaller than a proton to roughly a meter across.

There's some observational support for inflation, but still no good theory for what caused it.

An intriguing paper published in 2010 tried to remove the need for inflation by extending gravity theory to a new type of fundamental particles. The theory invoked a repulsive force called torsion. Torsion isn't noticeable at normal densities and temperatures, but in the conditions at the time of the big bang it would have allowed a universe to form from within a black hole. Our universe, then, would be space-time spawned by a black hole.[46] This idea has the side benefit of explaining the arrow of time. Time flows forward for us because of the time-asymmetric flow of matter into the event horizon from a parent universe. That is, on the other side of the event horizon, in the parent universe, time flows in the reverse direction. This wild situation arises because the events that have occurred since the big bang play out in reverse in the parent universe.

An even wilder theory published in 2014 reached into the toolkit of string theory. In an attempt to avoid the big bang singularity, researchers at the Perimeter Institute in Waterloo, Canada, proposed a theory that our universe arose as a result of the formation of a black hole in a higher-dimension universe. In our three-dimensional universe, black holes have event horizons that are two-dimensional. In a four-dimensional universe, a black hole would have a three-dimensional event horizon. Niayesh Afshordi and his colleagues propose that our universe came into being when a star in a four-dimensional universe collapsed into a black hole. The big bang is a mirage, a tracer of a higher-dimension event. They refer to Plato's allegory of the cave to describe the situation: "Two-dimensional shadows are the only things that the prisoners have ever seen—their only reality. Their shackles have prevented them from seeing the true world, a realm with one additional dimension to the world they know. . . . Plato's prisoners didn't understand the powers behind the Sun just as we don't understand the four-dimensional bulk universe."[47]

Making Black Holes in the Lab

Let's bring black holes down to Earth by asking this question: do we have the power to make a black hole? Before answering the question, let's recall just how extraordinary a black hole is. Schwarzschild radius is proportional to mass. To turn the Sun into a black hole it would have to be squashed down to a radius of 3 kilometers, corresponding to a density of 20 trillion kilograms per cubic meter. Turning the Earth into a black hole would mean squashing it to a radius of 9 millimeters, a little smaller than a ping-pong ball. That is an amazing density of 10^{24} kilograms per cubic meter. To put this in perspective, the density of a typical rock is 2,000 kilograms per cubic meter. With his fantastic strength, Superman can crush a lump of coal into a diamond, but that's just increasing the density from 900 to 3,500 kilograms per cubic meter. To reach black hole density you'd have to compress matter by another factor of 1,000 billion billion! Try that, Superman.

Black hole creation is far beyond our current capabilities. The Large Hadron Collider creates unprecedented energies, but it is a factor of 10 million times too weak to create black holes, even in theory (Figure 43).[48] That didn't stop news outlets from calling it the "doomsday machine" and speculating that it might create microscopic black holes that would sink to the center of the Earth and consume the planet. A search for microscopic black holes failed,[49] and the various doomsday scenarios were convincingly debunked.[50]

If extra dimensions exist, gravity in our universe may flow into other dimensions. That would explain why gravity is such a weak force. Also, since the energy required to make a microscopic black hole depends on the number of dimensions space has, it would be easier to create microscopic black holes. Looked at in this way, the fact that particle accelerators can't create miniature black holes is a refutation of extra dimensions. Also, energies sufficient to create microscopic black holes—

far beyond the capabilities of the Large Hadron Collider—are seen in cosmic rays from space every few months. Yet there's no evidence that cosmic rays create black holes. Finally, even if the collider could create black holes, they'd be so tiny, 10^{-23} kilograms, that they would require 3 trillion years to consume enough matter to grow to a kilogram. But if black hole theory is correct, they'd never get a chance to grow because they'd fizz away to nothing by Hawking radiation in a tiny fraction of a second.[51]

If miniature black holes *could* ever be made, they would offer a compelling means to travel to the stars. Interstellar travel is stuck in the starting blocks because we use chemical energy for our rockets. This inefficient fuel is adequate for getting people into Earth orbit and pay-

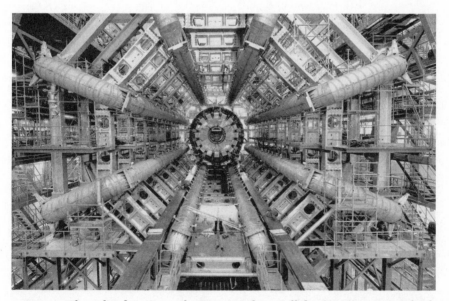

FIGURE 43. The Atlas detector at the Large Hadron Collider (LHC) in Switzerland. Eight toroidal magnets surround the detector where protons collide at incredible energies and speed close to the speed of light. Although matter is instantaneously compressed in the LHC, the densities are far short of what would be needed to create a black hole, and even if they were somehow sufficient, the resulting black hole would be so small that it would evaporate in a tiny fraction of a second by means of Hawking radiation. *M. Brice/Atlas Experiment © 2018 CERN*

loads around the Solar System, but hopeless for traveling the trillions of miles to even the nearest stars. However, the energy emitted by Hawking radiation from a microscopic black hole could propel a starship to a significant fraction of the speed of light. A black hole used for space travel would have to be small enough to manufacture, have a mass similar to that of a starship, and live long enough to be useful. A black hole weighing half a million tons would fit the bill. It would have a size of 10^{-18} meters, a power output of 10^{17} watts, a lifetime of three or four years, and, assuming 10% conversion to kinetic energy, would accelerate a starship to 10% of the speed of light in 200 days.[52] The black hole would be located at the focal point of a parabolic reflector to create forward thrust. That's the concept. The rest is just engineering.

6.

BLACK HOLES AS TESTS OF GRAVITY

NEWTON'S LAW OF gravity is only an approximation of the deeper level of reality described by Einstein's theory of general relativity. When the force of gravity is strong, the bizarre behavior of curved space-time manifests. Light bends, clocks run slow, and intuition fails us. A century after it was first published, Einstein's theory has passed all of its tests with flying colors, but almost all of the tests have been situations of weak gravity.

A black hole is the ultimate proving ground for general relativity. In a black hole, the distortions of space and time are extreme. At the event horizon, time is predicted to freeze. At the photon sphere, a distance 50% farther from the singularity than the event horizon, photons are predicted to orbit like satellites orbit the Earth. Gravity this strong cannot be created in any lab on Earth. Ideally we could run tests on a black hole fairly close to the Earth. However, the nearest stellar-mass black holes are hundreds of light years away and the nearest supermassive black holes are millions of light years away. Therefore, astronomers must use distant black holes to devise experiments to test gravity in new ways.

Gravity from Newton to Einstein and Beyond

Black holes can only be understood with Einstein's theory of gravity, but they're not the reason a new theory of gravity was needed. That story starts in England in 1665. At the age of twenty-three, Isaac Newton had already failed as a farmer, so his mother had sent him to Cambridge to study. The university was shut down due to plague, and Newton was forced to stay home, where he thought about gravity. Twirling a rock on the end of a string, he could see that the rock wanted to fly outward but the string provided a counteracting force. So what was the counteracting force that kept the Moon orbiting the Earth and the planets orbiting the Sun? By 1687, he had deduced the answer: a force diminishing with the inverse square of the distance. Newton detailed the theory of gravity in his masterwork, *Principia Mathematica*.

Astronomers were soon using the law to make increasingly accurate predictions. The comet that bears Edmund Halley's name was predicted to return in April 1759, and when it did Newton's reputation was burnished. A century later, French astronomer Urbain Jean Joseph Le Verrier was pursuing an anomaly in the orbit of Uranus, the first new planet to be discovered since antiquity. He deduced that it was perturbed by something exterior to its orbit, and he predicted the mass and position of the interloper. Neptune was discovered almost immediately at the Berlin Observatory. It seemed there was no limit to the explanatory power of Newton's theory.[1]

But there was a small dark cloud hanging in the blue sky: the orbit of Mercury. Mercury has a highly elongated orbit, and its closest approach to the Sun—the perihelion—shifts by 5,600 arc seconds per century as seen from Earth (about one and a half Moon diameters). Le Verrier's best calculations showed that the known planets and Newton's law could only explain a precession of 5,557 arc seconds. Such was the confidence in Newton's theory that an undiscovered interior planet called

Vulcan was hypothesized to explain the tiny difference.[2] Le Verrier died believing that Vulcan would be found, but it never was. In fact, Newton's theory was flawed.

In 1907, Einstein was just two years removed from his "miracle year," when he redefined physics, but he wasn't trying to improve on Newton's law of gravity. He was working in the patent office in Bern with plenty of time on his hands. Then he was startled by his "happiest thought": that a person in free fall would not feel their weight. This notion impelled him toward a radical new way of thinking about gravity.

Eight years later Einstein was in turmoil. He'd done most of his early work alone. Academia belatedly embraced him, and he had become a professor of physics in Prague, but it was an uneasy situation; anti-Semitism was on the rise in Europe and Einstein experienced it directly. It may be hard to believe, but Einstein struggled with the math of general relativity. He was at his most comfortable relying on his extraordinary physical intuition. For years he sketched out versions of the theory, but there were always flaws and omissions. In the summer of 1915 he gave a series of lectures on relativity at the University of Göttingen, and in November 1915 he made a breakthrough, which he presented in his fourth lecture at the Prussian Academy of Sciences, titled "The Field Equations of Gravitation." His crucial test of the equations was whether they could account for the anomalous shift in Mercury's orbit. The theory predicted an effect of 43 arc seconds per century—exactly the difference between what was observed and what Newton's theory predicted. "I was beside myself with joy and excitement for days," Einstein told a colleague. "The results of Mercury's perihelion movement fill me with great satisfaction. How helpful to us is astronomy's pedantic accuracy, which I used to secretly ridicule!"[3]

In Newton's theory, the source of gravity is mass. In Einstein's theory, mass is part of a more general quantity called the energy-momentum tensor. Think of a tensor as a fancy version of a vector, with information about a physical quantity at every position in space. Mass in general relativity is defined in curved space-time and has energy and momentum in each of three directions, so in Einstein's theory it takes ten equations to describe the relationship between mass and space-time. That's as far as we can go without joining the Mad Hatter and jumping down the rabbit hole of coupled, second-order, partial differential equations.

General relativity was just one of the foundational physical theories of the early twentieth century. The other was quantum mechanics, which explains the behavior of atoms and subatomic particles. These theories of the large and the small are incompatible. Relativity is "smooth" because events and space are continuous and deterministic. Everything that happens has an identifiable, local cause. Quantum mechanics is "grainy" because changes occur discretely by quantum leaps, and outcomes are probabilistic rather than definite. The most bizarre example of dissonance between the theories is quantum entanglement, in which properties of particles can be coupled instantaneously over large distances.[4] Einstein derided this as "spooky action at a distance" and was convinced there was a deeper theory of nature that would remove the weirdness of quantum mechanics.

He failed in his quest. Despite many attempts, Einstein couldn't find fatal flaws, or even poke significant holes, in quantum theory. He tried to generalize his geometric theory of gravity to include electromagnetism, leading him to become frustrated and increasingly isolated in this research. When he died in Princeton in 1955, he left an unsolved set of equations on his blackboard.

The mantle, or perhaps the burden, of reconciling these two great theories has been taken up by succeeding generations of physicists. The ultimate goal is a "theory of everything" that will explain all physical phenomena. There are four fundamental forces in nature. Two apply on a subatomic scale: the strong and weak nuclear forces. Two apply over very large distances: electromagnetism and gravity. Physicists went some way toward unifying these forces in the second half of the twentieth century. Accelerator experiments in the 1970s showed that electromagnetism and the weak force that's responsible for radioactivity are manifestations of one electro-weak force. Additional experiments have almost succeeded in bringing the strong nuclear force into the mix. This edifice is called the standard model of particle physics.[5] But gravity stubbornly resists being part of this model. Nobody has ever seen a graviton, the hypothetical particle that carries the gravity force. Unification that brings in gravity may only occur when the temperature is a fantastic 10^{32} Kelvin (Figure 44). And the only situation we know of with that temperature is 10^{-43} seconds after the big bang, when the universe was the size of a fundamental particle and general relativity crashes and burns in the initial singularity.

There are several approaches to quantum gravity.[6] Quantum loop gravity follows the thought process of Pythagoras, who imagined cutting a stone in half and half and half again until a limit was reached. In this

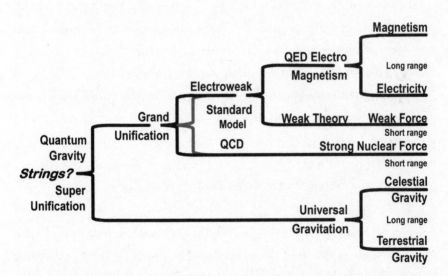

FIGURE 44. The four fundamental forces of nature consist of two with infinite range, gravity and electromagnetism, and two that operate on subatomic scales, the strong and weak nuclear forces. They all have very different strengths but there is evidence that at extremely high energies they unite into a "super-force." The unification of weak and electromagnetic forces was seen at accelerators in the 1970s, and there are hints of a "grand unification" with the strong force. *CERN/CMS Collaboration*

case, an inch is divided in half and half and half again until "atoms," or indivisible units of space, are reached. Quantum loop gravity is an attempt to directly extend the formalism of quantum mechanics to the gravity force. The more radical approaches involve string theory and extra dimensions of space beyond the familiar three. Going from New-ton to Einstein to beyond—from rigid and linear to supple and curved to evanescent and grainy—is the most important unfinished project in physics. Progress has been slow and the work is fiendishly difficult.

We saw in chapter 1 that black holes are not only situations of extreme gravity but also situations where quantum effects are impor-tant. Any new theory that reconciles the "smooth" world of curved space-time with the "grainy" world of subatomic particles will meet its most important challenge in black holes.

Einstein once said there are only two things that might be infinite: the universe and human stupidity. And he wasn't sure about the uni-verse.[7] Some of the smartest humans on the planet are trying to come up with a theory of quantum gravity. They might succeed, but they might not. Meanwhile, progress can be made by testing and trying to break general relativity. As another great physicist, Richard Feynman, said, "We are trying to prove ourselves wrong as quickly as possible, because only in that way can we find progress."[8]

What Black Holes Do to Space-Time

A black hole can be defined as a region of space-time so curved that it is "pinched off" from the rest of the universe. But even at some distance from the black hole, space-time curvature will cause particles and light to be deflected. When Einstein developed general relativity, no black holes were known. So his theory was tested with a much more subtle effect: the slight deflection of light from a distant star as it grazes the Sun on its way to the Earth. This is most easily observed during a solar eclipse, when the Sun is blotted out by the Moon and the background star is visible.[9] In 1919, just three years after the theory of general relativity was published, Arthur Eddington and his colleagues measured the deflection simultaneously from Brazil and South Africa. It matched Einstein's prediction.[10]

The result made the front page of most newspapers. The drama was undoubtedly heightened by the symbolism of a British scientist confirming the work of a German scientist at the end of a long and bloody war. Einstein became a celebrity overnight. He was supremely confident about the outcome. Asked what his reaction would have been had general relativity not been confirmed by the expedition, he said, "Then I would feel sorry for the dear Lord. The theory is correct anyway."[11]

Mass bends light. Given the importance of this fact for Einstein's theory and reputation, it's surprising that he was slow to recognize the broader implications. He knew that if light rays passed close to a sufficiently massive object, they could be bent enough to converge and produce a magnified image, or multiple images, of a background source. As the process resembles the bending of light through a lens, researchers called it gravitational lensing. At the urging of an engineer colleague, Einstein finally published a paper on lensing in 1936, with this strikingly diffident preface: "Some time ago, R. W. Mandl paid me a visit and asked me to publish the results of a little calculation which I had made at his request. This note complies with his wish."[12] He wrote a self-deprecating note to the editor of the journal: "Let me also thank you for your cooperation with the little publication, which Mister Mandl squeezed out of me. It is of little value, but it makes the poor guy happy."[13]

Einstein was dead wrong about the value of gravitational lensing. It has become an essential tool in modern astrophysics. It has been used to map out the dark matter in galaxies and across the universe, to measure the geometry and expansion rate of the universe, to constrain dark energy, to do surveys for brown dwarfs and white dwarfs, and to detect exoplanets smaller than the Earth (Figure 45).

Einstein thought the lensing effect would be too small to be mea-

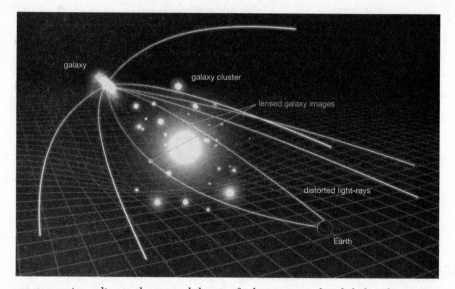

FIGURE 45. According to the general theory of relativity, mass bends light. If a massive object like a cluster of galaxies lies between us and a more distant galaxy, space-time is distorted and the light from the distant galaxy is bent around the cluster. This causes distorted and magnified images to form. Since the lensing is caused by all mass, not just visible matter, this is a way to measure the amount of dark matter in the universe. *L. Calcada/NASA/ESA*

surable. But within months of his paper, Caltech astronomer Fritz Zwicky realized that billions of stars combined within galaxies could produce observable lensing. In a prescient paper, he outlined essentially all the modern uses of gravitational lensing.[14] However, it took until 1979—more than forty years—for lensing to be observed. The tool was a supermassive black hole billions of light years away.

A group of researchers led by British radio astronomer Dennis Walsh found two quasars with identical spectra using the 2.1-meter telescope at Kitt Peak. The odds of finding two quasars with identical spectra so close on the sky were very low—so low that on his way to Kitt Peak, Walsh wrote a bet on the blackboard of his colleague Derek Wills: "No QSO's, I pay Derek 25 cents. One QSO, he pays me 25 cents. Two QSO's, he pays me a dollar." Walsh recalled, "When I called Derek the next morning and told him what we had found we laughed and I said

'You owe me a dollar. Suppose I'd said to you 'Two QSO's, same red-shift, $100' would you have taken it?' He said 'Of course.' So I lost $99 and kept a friend. . . . I had four teenage sons, none particularly inter-ested in science. So when they asked me 'What good is this gravitational lens' I was able to say 'Well, I made money out of it.' "[15]

The quasars looked like identical twins. But rather than two qua-sars that just happened to have the same spectra, they were more like a mirage. Light from one quasar took two different paths around an intervening galaxy, giving two images. A massive galaxy bends light very subtly, by only a thousandth of a degree. In this first gravitational lens, light travels for 8.7 billion years to reach us, but it travels just over a light year farther going past one side of the galaxy compared to the other. Since the light from the quasar varies in brightness, there is a time delay of just over a year in the variations seen in one image compared to the other. This has been used for a clever measurement of the expansion rate of the universe.[16]

Gravitational lensing is rare because it depends on a near-perfect alignment of a background quasar and a foreground galaxy. With many thousands of quasars studied, less than 100 cases of lensing have been found. In a dozen of them, the alignment is perfect, so instead of mul-tiple images, the intervening galaxy turns the quasar point source into an Einstein ring[17]— an exquisite display of general relativity in action. Depending on the geometry, light from accretion energy near a super-massive black hole appears as an arc, multiple images, or a perfect ring.

When the Hubble Space Telescope started work in the 1990s, another type of lensing situation was discovered. Instead of light from a single quasar being lensed into multiple images, light from numerous distant galaxies is lensed by an intervening galaxy cluster. Sometimes multiple images are formed, but more often the background galaxy light is sheared into an arc. The signature of this type of lensing is a cluster surrounded by little arcs arranged in concentric circles around the center of the clus-ter (Figure 46). Each distorted image is an experiment in gravitational

FIGURE 46. Gravitational lensing by clusters of galaxies was predicted by Fritz Zwicky in 1937 but was not observed until astronomers gained the sharp imaging capability of the Hubble Space Telescope in the 1980s. In this image, Abell 2218 is causing the distortion and magnification of many more distant galaxies. The lensed arcs form concentric circles around the center of mass of the cluster. In some lensing situations, one distant galaxy can have five or seven separate images. *W. Couch, R. Ellis/NASA/ESA*

optics. Several hundred clusters have shown these arcs, so astronomers have accumulated tens of thousands of examples of mass bending light.[18]

All mass bends light, whether it's visible or not, so lensing is the best tool astronomers have for mapping dark matter in galaxies, in clusters, and in the space between galaxies. Lensing provides the best evidence that dark matter exists and is a dominant and pervasive component of the universe.

How Black Holes Affect Radiation

The event horizon of a black hole is a place where time stands still and radiation stands still. It's a premise of Einstein's special relativity that light has a universal and constant speed of 300,000 kilometers per second. Light leaving a black hole struggles against gravity so intense that its speed is quelled, its energy sapped. The effect is called gravitational redshift. The event horizon of a black hole corresponds to a place where the redshift is infinite and light is trapped.

Without a black hole to test the theory on, how can we understand what gravity does to radiation? Let's use a thought experiment situated on Earth. Imagine a tower where we send a photon from bottom to top, turn its energy into mass (according to $E = mc^2$), let the mass drop to the bottom of the tower, and turn it back into a photon. Sounds straightforward. But wait. If we drop the mass, it picks up speed and gains gravitational energy. The amount gained is mgh, where m is the mass, g is the acceleration due to the Earth's gravity, and h is the height of the tower. When we turn the mass back into a photon it has more energy. We could keep doing this over and over and create energy and get rich! Since nobody has made money by cycling light up and down, there must be a flaw in our assumptions. The only way to conserve energy—in other words, keep the energy the same—in this scenario is to posit that light is affected by gravity in the sense that it loses energy as it climbs away from the Earth's surface. Losing energy means that the light shifts to a longer, or redder, wavelength. This is gravitational redshift.

Imagine a clock based on the frequency of light. Put the clock at the bottom of the tower. If we observe from the top, the photons lose energy reaching us, so they decrease in frequency. We see the clock running slower. Conversely, if we're at the bottom of the tower looking up, a clock at the top will run slightly faster. Time passing more slowly in strong gravity is another prediction of general relativity. An amusing example, attributed to the physicist Richard Feynman, is the prediction that the center of the Earth is two and a half years younger than the surface.[19] It's called gravitational time dilation. The redshift and time dilation effects are closely related. Light and other forms of electromagnetic radiation have a wavelength that's inversely proportional to frequency. As light's energy is diminished in its struggle against gravity, its wavelength gets longer or redder and its frequency gets smaller, which is the same as saying that light's "clock" is running more slowly.[20]

The first observation of gravitational redshift was made by Walter Adams in 1925. He measured the shift in the spectral lines of the nearby white dwarf Sirius B. As it is part of a binary system, its mass is known, and the shift is a few parts in 10,000 in wavelength, compared to a few parts in a million for a less compact star like the Sun. Unfortunately, the measurement was flawed by light contamination from the far brighter companion Sirius A, so scientists didn't consider the effect confirmed.

The first laboratory test of general relativity was an experiment by Robert Pound and his graduate student Glen Rebka in 1959. They measured the spectral shift in gamma rays from radioactive iron traveling 22.5 meters up a tower on the Harvard campus. The tiny loss in energy, less than 3 parts in 10^{15}, confirmed the prediction of general relativity at the 10% level (Figure 47).[21] Improvements came with the use of atomic clocks as probes of gravity. In 1971, a cesium atomic clock flown at high altitude on a commercial jet gained 273 nanoseconds compared to an identical clock at the U.S. Naval Observatory,[22] and in 1980, a better test used a maser clock flown on a rocket to improve the agreement with relativity to 0.007%.[23] The current state of the art measures quantum interference of atoms. General relativity has been confirmed with the phenomenal accuracy of less than a millionth of a percent.[24] We can show that a clock actually runs faster when we raise it less than a meter!

Astronomers have got in on the act as well. Clusters of galaxies are the most massive objects in the universe. Photons from the center of the cluster, where there are many galaxies, should lose more energy than photons from the edge, where there are fewer. A group at the Niels Bohr Institute led by Radek Wojtak looked for this effect, which is so small that they had to combine data from 8,000 clusters to detect it.[25] Once again, Einstein's theory was confirmed.

A good thought experiment elicits the reaction: of course, that's obvious! Recall the English biologist Thomas Huxley's reaction when he

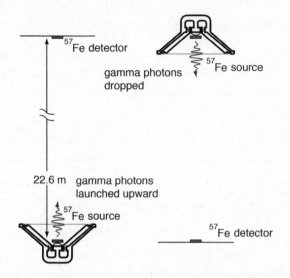

FIGURE 47. The first experimental test of general relativity in 1959 was the most precise physics experiment ever attempted at the time. Harvard physicists Robert Pound and Glen Rebka measured the energy of gamma rays from an iron-57 radioactive decay traveling upward and downward over a distance of 22.6 meters. The photon traveling downward was blueshifted and the photon traveling upward was redshifted by exactly the amount predicted by general relativity. The experimental precision required for this test was a few parts in 10^{15}. *R. Nave/Hyperphysics*

heard about Darwin's theory of natural selection: "How extremely stupid not to have thought of that!"[26] In Einstein's elevators, the beauty of general relativity is revealed. An elevator free-falling toward the ground is the same as an elevator drifting in deep space, because the gravity force has been removed. And an elevator being accelerated through space at 9.8 meters per second is the same as an elevator on the ground, because acceleration by gravity can't be distinguished from acceleration by any other force. In the second case, imagine you shine a flashlight across the elevator. During the instant of time that it takes for light to reach the other side the elevator is accelerating, so the light travels a downward curving path across the elevator. Einstein's theory says the same thing must happen with the stationary elevator on the ground.

Light "falls" due to gravity. Or, in the language of relativity, the mass of the Earth curves space and light bends slightly as it follows the curved space-time near the Earth.

So far we have described the "classical tests" of general relativity. They use situations where gravity is so weak that space-time curvature and distortion is slight and extremely accurate measurements are needed. Nearly fifty years ago, Irwin Shapiro, the longtime director of the Harvard–Smithsonian Center for Astrophysics, proposed an ingenious weak gravity test of the theory. He realized there would be a slight delay in the round-trip travel time of radar signals bouncing off other planets if the path of the photon took it near the Sun. Using measurements of radar bounced off Mercury and Venus, before and after they were eclipsed by the Sun, he confirmed general relativity at a 5% level.[27] This test was repeated in the outer Solar System by NASA's Cassini spacecraft to give agreement at the 0.002% level.[28]

These tests affirm general relativity and its superiority over Newton's theory. But there's something vaguely unsatisfying about testing relativity in places where space is as flat as an Iowa cornfield. It's like test-driving a Lamborghini in a parking lot. Sure, it performs better than your old Ford Taurus, but that's setting the bar low. It's much better to drive both cars fast in the mountains, where the Lamborghini powers up the hills and hugs the curves while the Taurus overheats and careens off the road. Astronomers look forward to eventually testing the theory with black holes, where the effects on radiation should be spectacular. As we'll see in the next section, large gravitational redshifts have been detected from black holes using spectroscopy of the accretion disk.

Inside the Iron Curtain

The vicinity of a black hole is the ultimate test for general relativity. How close can we get with observations? The event horizon, through which no information can reach us, defines the limit. General relativity also describes several important scales beyond the event horizon. The first is called the photon sphere, where light is trapped and travels on circular orbits around the black hole. Since mass bends light, we can think of mass bending light into a circle. If you could go there, a photon might begin at the back of your head, orbit the black hole, and enter your eye, allowing you to see the back of your head. For a stationary black hole, the photon sphere has a radius 1.5 times the Schwarzschild radius.[29] A rotating black hole has two photon spheres, and drags space with it as it spins. The inner photon sphere moves in the direction of rotation and the outer photon sphere moves against the rotation. Think of a swimmer trying to escape the maelstrom. They hold their ground by swimming against the current; if they swim with the current they are pulled closer to their doom. Since the photons are trapped, a photon sphere has never been observed.

We enter the realm of observation at the inner edge of the accretion disk. As particles are tugged toward the black hole by gravity, they rub against one another, so the accretion disk is a plasma whose temperature decreases going out. The inside edge is defined by an innermost stable orbit that is 3 times the Schwarzschild radius for a nonrotating black hole and a little bit beyond the event horizon for a rapidly spinning black hole.[30] Inside this stable orbit, a particle plunges into the black hole and disappears forever. The inner edge of the accretion disk for a low-mass black hole has a temperature of 10 million Kelvin, and for a supermassive black hole it has a temperature of 100,000 Kelvin. Gas this hot emits copious amounts of X-rays.

Can we see the inner edge of the accretion disk? No. The angular scale is far too small for any telescope to resolve. A nearby black hole at a distance of 100 light years has an inner edge that subtends an angle of 10^{-9} arc seconds. That's like trying to see the head of a pin on the surface of Mars. The situation improves slightly for a supermassive black hole like the inactive ones that have been found in the centers of nearby galaxies. They are several million times farther away but have event horizons a billion times bigger, so their inner accretion disks subtend an angle of 10^{-7} up to 10^{-6} arc seconds. That's a few hundred times smaller than the resolution of the radio interferometers described earlier, so still beyond the reach of observational astronomy.

The only way astronomers can peer inside the iron curtain is to use spectroscopy. The gas in an accretion disk is overwhelmingly made of ions of hydrogen and helium, but two of every million particles are ions of iron. The region just beyond the accretion disk is an extremely hot corona. X-rays from the corona irradiate the slightly colder accretion disk and their energies are just right to excite spectral transitions of iron. Even though iron is a rare element, these spectral features are sharp and strong. An X-ray spectrum shows how the gas is moving, because the approaching part of the accretion disk is blueshifted while the receding part is redshifted. X-rays from the inner part of the accretion disk also suffer a strong gravitational redshift, so the spectral line of iron is broadened and also skewed to lower energies (Figure 48). X-rays offer the exciting possibility of measuring gravity within spitting distance of the event horizon.[31]

These observations were made possible by the launch of the X-ray satellite ASCA in 1993. The first detection of X-rays from the inner edge of a massive black hole accretion disk followed a year later.[32] Gravitational redshift of X-ray spectral lines has now been seen from a dozen stellar-mass black holes and a similar number of supermassive black holes. Then, a puzzling X-ray phenomenon discovered years earlier became a second window onto the vicinity of a black hole.

X-Rays Flickering Near the Abyss

In the 1980s, when X-ray satellites began monitoring compact stars and star remnants, sources whose X-rays varied rapidly were seen. The flick-

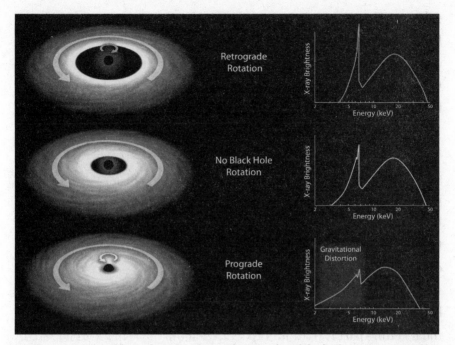

FIGURE 48. A spectral line of iron can be used as a probe of the high-temperature inner region of the accretion disk surrounding a black hole. The line is skewed to low energy by gravitational redshift from the black hole. The inner edge is far from the black hole when the black hole rotates in the opposite direction to the accretion disk (retrograde), and closer when it rotates in the same direction (prograde). These differences are seen in the X-ray spectrum. *NASA/JPL/California Institute of Technology*

ering wasn't rhythmic, so the phenomenon was called quasi-periodic oscillation. Oscillations were first seen in white dwarfs, and later in neutron stars and black holes.

It took a while for astronomers to untangle the astrophysics behind these variations. In different sources the timescales ranged from a second to as little as a millisecond, and the periodic behavior was often lost in the noise of more chaotic variations. Black holes showed a particular pattern of brightening and dimming, at first taking ten seconds to complete an oscillation, then weeks or months later speeding up to a tenth of a second before the variations stopped, after which the cycle repeated. Observations and models of the archetypal black hole Cygnus

X-1 revealed the source of the variations. They are pulses left by gas as it leaves the inner part of the accretion disk and plunges toward the event horizon. It's thrilling to see the death throes of material falling into a black hole in real time.[33]

Astronomers suspected that the frequency of variations might depend on the mass of the black hole. Gas spirals inward in the accretion disk, moving faster and faster, and it piles up near the black hole, releasing a torrent of X-rays. This congestion zone is close in for small black holes, so the X-ray "clock" ticks quickly. It's farther out for larger black holes, so the X-ray "clock" ticks more slowly. The behavior is so reliable that the X-ray variations have been used to measure black hole mass,[34] including the smallest known black hole. Just 15 miles across and 3.8 times the Sun's mass, it's barely above the mass limit of a neutron star.

Recently, a group led by Adam Ingram at the University of Amsterdam combined X-ray data on variability and the shape of the iron line. He started working on quasi-periodic oscillations for his PhD in 2009 and says, "It was immediately recognized to be something fascinating because it is coming from very close to the black hole." Using data from two X-ray satellites, his group showed that the orbiting material was caught in a gravitational vortex created by the black hole: "It's a bit like twisting a spoon in honey. Imagine that the honey is space and anything embedded in the honey will be 'dragged' around by the twisting spoon." They chose a black hole with an oscillation time of four seconds and observed it carefully for nearly three months. The iron line showed exactly the behavior expected from general relativity. "We are directly measuring the motion of matter in a strong gravitational field near to a black hole," said Ingram.[35] It's still one of the very few tests of Einstein's theory in this regime.[36]

Quasi-periodic oscillations have also been seen in active galaxies. The timescales of variation are hours to months rather than seconds.[37] The exciting implication is that accretion disks behave in similar ways over a huge range of physical scales, from stellar black holes to supermassive black holes in distant galaxies.

When a Black Hole Eats a Star

What happens when a supermassive black hole eats a star? In 1998, Martin Rees ventured an answer. He had been thinking for years about how it might be possible to detect the dark black holes that should lurk at the center of every galaxy. He thought about what might befall any unfortunate star that ventures into a region of extreme gravity. As it gets close to the black hole, the star is first stretched and then ripped apart by tidal forces. Some of the debris is ejected at high speed and the rest is swallowed by the hole, causing a bright flare that might last for several years.[38]

Stars avoid this fate unless they travel very close to a black hole. Every black hole has a tidal disruption radius. Outside this limit stars maintain their shape. Once a star enters this space the destruction begins. About half of the star's mass is ejected, and the other half moves onto elliptical orbits that gradually deliver the gas into an accretion disk. The black hole feeds on this material just outside the event horizon and the conversion of gravitational energy into radiation makes a bright flare.[39] Sometimes the event triggers relativistic jets (Figure 49). Imagine the Sun approaching the black hole at the center of our galaxy. Nothing would happen until the Sun was within 100 million miles of the event horizon; then the Sun would be torn apart and all the planets, including the Earth, would be scattered like tenpins, with an equal probability of being ejected to safety or consumed by the black hole. Approach that close is unlikely, so tidal disruption events are rare, occurring about once every 100,000 years per galaxy.

For a Sun-like star approaching a central black hole of a few million solar masses, the radius of tidal disruption is far outside the Schwarzschild radius. But the Schwarzschild radius grows linearly with mass while the disruption radius grows more slowly, so black holes bigger than 100 million solar masses consume stars before they can be

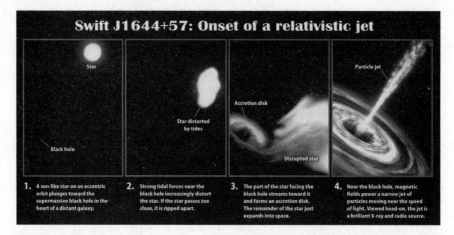

FIGURE 49. The tidal disruption of a single star by a massive black hole in a distant galaxy leads to a flare seen by a NASA satellite. The star is on an eccentric orbit so it passes near the black hole and is ripped apart by strong tidal forces. Some of the gas feeds an accretion disk and some is lost from the gravitational influence of the black hole. The accretion disk forms jets which accelerate high-energy particles, which then emit copious amounts of radiation toward the Earth. *NASA/Goddard Space Flight Center/Swift*

ripped apart. Think of big black holes swallowing the carcass whole, while smaller black holes tear the meat off to eat it. Also, the fate of a star depends on its size and stage of evolution. Large stars suffer stronger tidal forces. So a red giant heading into the galactic center would get disrupted much farther out than the Sun would, while a white dwarf would disappear inside the event horizon without being disrupted. Numerical simulations suggest that the accretion rate after a disruption event is sensitive to the mass of the black hole. If the simulations can be trusted, the time between disruption and peak flare brightness might be used to "weigh" a black hole. For a star like the Sun, that time delay is a month for a black hole 10^6 times the mass of the Sun, and the time delay increases to three years for a black hole 10^9 times the mass of the Sun.

What do the observations say? About twenty tidal disruption events have been seen with X-ray telescopes, including a couple where the accretion is so efficient that the brightness far exceeds the limit defined by Eddington a century ago.[40] A small set of events showed that a surge in accretion can power the relativistic jets that are seen in radio quasars.[41] All of these examples are in distant galaxies, so astronomers got very excited when they realized that a gas cloud called G2 was heading for the black hole in the center of our galaxy. In late 2013, the gas cloud passed very close to the massive black hole and . . . nothing. But a year or so after that close passage, the rate of X-ray flares increased by a factor of 10, to one a day. This has led to speculation that G2 was not a gas cloud but a star with a large envelope, so that it would have taken longer for material to be ripped off and descend into the black hole.[42] The show isn't over. After fifteen years of gathering data, X-ray astronomers are waiting for G2 to make another pass. The anticipation is only slightly dimmed by the fact that everything we are watching in the galactic center happened 27,000 years ago.

Meanwhile, optical astronomers have their eyes glued to S2, a star that loops around the galactic center black hole every sixteen years. They have a new tool called GRAVITY that combines light from the four 8.2-meter telescopes of ESO's Very Large Telescope to give the angular precision of a single telescope 130 meters across. In 2018, S2 will pass very close to the black hole and provide an unprecedented chance to test general relativity. It's expected to pass just 17 light-hours from the event horizon, traveling at 3% of the speed of light. It may be ripped apart or devoured entirely.

The destruction of a star by a black hole certainly captures the imagination. In 2015, it led to a news story pushing a culinary analogy: "Black holes devour stars in gulps and nibbles."[43] That spawned this overheated headline in England's *Daily Mail* newspaper: "Echoes of a stellar massacre: Gasps of dying stars as they are torn apart by supermassive black holes are detected."[44] Apart from the facts that stars don't have feelings, they don't make sounds, and sounds can't travel through a vacuum, the headline is quite accurate.

Taking a Black Hole for a Spin

Black holes are remarkably simple; the "no hair" theorem says they're described by just two numbers, mass and spin. We've talked in the first part of the book about ways to measure black hole mass, which usually involve an orbit with a visible companion, if the black hole is a collapsed star, or its effect on the motion of nearby stars, if it is massive and at the center of a galaxy. But what about spin?

Gravity doesn't depend on rotation in Newton's theory. But in Einstein's theory, mass couples to the geometry of space-time. In 1918, it was predicted that rotation of a massive object would distort space-time, making the orbit of a smaller nearby object precess, like the pivoting of a spinning top. This twisting of the contours of space is called frame dragging. Recall Poe's vivid description of the maelstrom. As with other subtle effects of general relativity, the first place to look is close to home.

The Earth twists space-time as it spins, but the effect is so minute that for decades it was considered impossible to detect. In 2004, NASA launched a satellite called Gravity Probe B to measure space-time curvature caused by the Earth and the even more subtle frame dragging caused by its rotation. The tools for the job were four gyroscopes the size of ping-pong balls. Gyroscopes are often used to guide spacecraft; their rotation axes point in a fixed direction. The gyroscopes on Gravity Probe B contained quartz spheres coated in niobium. They're among the most accurately machined objects ever made, spherical to within forty atoms. If scaled to the size of the Earth, the highest peaks and valleys would be no larger than a person. They were shielded from their containers by a thin layer of liquid helium. At that temperature, the spheres became superconductors and the electric and magnetic fields they generated were used to keep them aligned.[45]

Gravity Probe B began its sixteen-month mission fifty years after it was initially funded.[46] The gyroscopes locked onto a bright star in the constellation of Pegasus. The satellite measured space-time curvature by the tiny angle the gyros "leaned into" the gravity of the Earth, and it measured frame dragging by the even tinier angle the gyros "lagged" the spinning Earth. Unexpected noise reduced the sensitivity of the experiment and slowed the analysis. These headaches meant that final results weren't published until 2011.[47] Einstein's predictions of space-time curvature were confirmed to within 0.5% and his predictions of frame dragging were confirmed to within 15% (Figure 50). When the dust settled, Gravity Probe B was shown to be a successful (although exhausting) technical tour de force.

Spin has a different implication for low- and high-mass black holes. Black holes in binary star systems are more massive than their companions, so their spin won't change much due to that interaction. Their spin

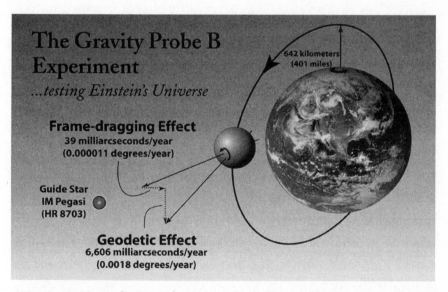

FIGURE 50. Gravity Probe B tested two particular predictions of general relativity in the weak field situation of Earth orbit. Gyroscopes were used to very accurately lock the satellite onto a celestial reference frame. The satellite measured geodetic precession, or the amount by which the gyroscopes "leaned into" the Earth's gravity, and it measured the frame-dragging effect where the gyroscopes "lagged" the spinning Earth. Both measurements matched the predictions of general relativity. *C. W. F. Everitt/Phys.Rev. Lett./American Physical Society*

rate is a direct relic of their formation in a supernova explosion. Massive black holes, by contrast, grow over cosmic time by consuming gas and stars in the inner parts of their galaxy, but also by mergers with black holes in other galaxies. So the spin of a massive black hole encodes the history of its growth through accretion and mergers. Therein lies the motivation to make this difficult measurement.

Spin has been measured for several dozen supermassive black holes. Most often, the measurement uses the shape of the iron spectral line when it is reflected off the inner edge of the accretion disk. From a million to a billion solar masses, most black holes spin between 50% and 95% of the speed of light.[48] Such rapid spin rates suggest the black holes grew after a single major merger with another galaxy, where most incoming material arrives from one direction, as opposed to being built by many minor mergers of material coming from different directions, which would average out to a slow spin rate.

The best way to measure spin is by using the types of data that probe the inner accretion zone: spectroscopy of the iron line, quasi-periodic oscillations, and the rare tidal disruption events.[49] What's the limit on the spin rate of compact stars? For neutron stars, the rotation rate can only be measured for the subset in which a hot spot emits radio waves that sweep across the sky like a searchlight. The fastest pulsar spins 716 times a second.[50] Theory suggests the limit is 1,500 times a second, above which the neutron star would break apart. The maximum spin rate for a black hole isn't set by the structure of matter, since all information is hidden by the event horizon. It's set by a spin rate where the circumference of the event horizon would be moving at the speed of light. GRS 1915+105, 35,000 light years away in the constellation Aquila, is spinning at a blinding 1,000 times per second. That's over 85% of the maximum rate. Archetypal black hole Cygnus X-1 doesn't spin quite as quickly, but its rate of 790 times a second is 95% of the theoretical limit.[51]

Let's try and visualize these whirling dervishes. GRS 1915+105 is 14 times the mass of the Sun, so it has a Schwarzschild radius of 42 kilometers. Imagine this black hole hovering in the upper atmosphere above London. It would be a dark blot covering a tenth of the sky, and it would cast a shadow not only on London but over much of southern England. Even though it's 300 times smaller than the Earth, it's far more massive than the Sun. The turbine of a military jet engine is spinning so fast that it emits a note two octaves above middle C, in the range of a soprano singer. If this black hole could make a noise, it would be at a similar pitch, even though the black hole is the size of a large city!

At the other extreme, let's think about the large member of the black hole binary in the active galaxy OJ 287, which is 3.5 billion light years away. The mass of this black hole is 18 billion times the mass of the Sun, its Schwarzschild radius is 50 billion kilometers, and it's spinning at 100,000 kilometers per second, or a third of the speed of light, at its equator.[52] This situation is harder to visualize, but let's imagine the supermassive black hole is lurking somewhere in space above the Solar System. It's 10 times the size of the Solar System but has the mass of a small galaxy. A black hole this size has a more leisurely spin rate, but it still manages to rotate once in five weeks. For comparison, to show how bizarre this behavior is, an object in the Solar System following Newton's laws and located at the same distance from the Sun as the event horizon of this black hole would only orbit once every 5,000 years. Nothing in the nearby universe prepares us for such extreme motion.

The Event Horizon Telescope

"We're swinging for the fences." Shep Doeleman is sipping coca leaf tea to combat the effects of altitude at the top of a 15,000-foot volcano in southern Mexico. Despite these optimistic words the night isn't going smoothly as he fights problems with the instrument and his radio telescope steadily fills with fresh snow. "If something is dancing around the edge of the black hole, it doesn't get any more fundamental than that. Hopefully, we'll find something amazing."[53]

Doeleman was a physics student at Reed College in Portland, Oregon, where science students ran their own nuclear reactor and the air in the Student Union was often thick with marijuana smoke. Wanderlust led him to take two years off before graduate school and spend most of that time doing scientific experiments in Antarctica. As a graduate student at MIT, he tried plasma physics and geology before settling on radio astronomy when he saw the beautiful maps of quasar jets produced by the technique of very long baseline interferometry. Doeleman realized this technique offered the best chance to take a picture of a black hole, and he knew exactly where to look: the ultra-compact radio source in the direction of Sagittarius called Sagittarius A*.

The center of our galaxy is a compelling target for this research. Apart from the fact that it has the most convincing evidence of any black hole candidate, it's also the easiest to study. The angle subtended by the event horizon of the black hole at the galactic center is 50 micro-arc seconds. That's a tiny angle, but it's 10 times easier to resolve than the event horizons of supermassive black holes in external galaxies and several thousand times easier to resolve than the event horizons of the nearest stellar mass black holes. So it has become the watering hole for astronomers wanting to probe a black hole and test general relativity in a new way.

Doeleman is the young leader of a project called the Event Horizon Telescope.[54] The Event Horizon Telescope isn't a single facility; it's an array of eleven radio telescopes spread around the world. Dishes from Chile to Antarctica to Hawaii to Arizona to Spain all act in concert to mimic the imaging sharpness of a single telescope the size of the Earth. Operating a telescope as big as the world requires atomic clocks accurate to a second in a century. Astronomers from twenty institutions are working on the project. Data is gathered at short radio wavelengths of a millimeter or less. Millimeter radio waves are affected by water vapor in the atmosphere, so most of the telescopes are in cold, dry locations. As a result, Doeleman has not only watched telescopes fill with snow, he's had to wear an oxygen mask to test equipment at an altitude of 16,000 feet in the Andes, and he's suffered risks to his extremities using a telescope at the South Pole.

A group of thirty scientists and engineers run the radio dish on Kitt Peak in southern Arizona that's a vital part of the array. My University of Arizona colleagues Feryal Ozel and Dimitrios Psaltis are using numerical relativity and ray tracing on a powerful supercomputer to calculate the appearance of the black hole. Another colleague, Dan Marrone, winters over in Antarctica every year to tend another antenna in the array, the South Pole Telescope. These scientists are in their forties, part of a generation that's determined to get to the bottom of black holes—at least metaphorically.

In the quest for higher and drier observing sites, the South Pole can't be beat. The dome of ice rises 8,500 feet above sea level and the humidity is less than 10%. All that water is frozen into ice as hard as granite underfoot. I hope to go someday, but think I'll forgo that endless winter night, when the gales howl and the temperature hovers around −60°C. If you winter over at the South Pole you have to be very confident of your sanity and that of your colleagues. Being a radio astronomer, Dan Marrone doesn't need a dark sky to detect millimeter waves, so he goes during the Antarctic summer, trading the warm Tucson winter for a temperature just below freezing, which is as warm as it gets on the bottom of the world. There's something poetic about going to a place of endless light to take a picture of endless darkness.

The project already has some impressive results and the array isn't even operating at full strength yet. Material is falling into the galactic center and that region should be very bright, given the size measured by the Event Horizon Telescope. However, it's faint, so energy must be disappearing into an event horizon, which is strong evidence for a black hole.[55] Early data show that the accretion disk is seen close to edge-on, which means that our perspective allows the disk rotation to be measured, thereby putting constraints on the spin of the black hole. The variability of the compact radio source is associated with changes in the accretion flow very close to the black hole. Simulations suggest the array will soon be sensitive enough to meet its design goal: to make the first image ever of a black hole (Figure 51).

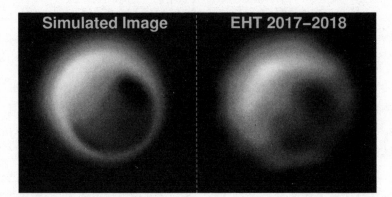

FIGURE 51. On the left is a simulated image of the black hole at the center of our galaxy. The simulation uses an accretion flow method and shows light lensed around the black hole into a distinctive ring encircling the black hole shadow. The ring diameter is 5 times the Schwarzschild radius. The image is bright on the approaching side and faint on the receding side of the accretion disk. On the right is an image with the expected performance of the Event Horizon Telescope in 2018. *A. Broderick, V. Fish/Perimeter Institute and University of Waterloo/ApJ, vol. 795, reproduced with permission/copyright AAS*

The image, if it can be obtained, will be a small, dark circle of nothing. General relativity says the shadow will be 50 million miles across, which, as seen from the Earth, is like the size of a poppy seed in New York as seen from Los Angeles. The silhouette will be doubled in size by the gravitational bending of light, and rimmed in light from surrounding stars. If the image isn't exactly circular it will be evidence against the "no hair" theorem of black holes.[56] But if the image has the shape and size predicted by relativity it will be the best visual evidence yet that space and time really can curl into a ball and 4 million suns can disappear with barely a trace.

7.

SEEING WITH GRAVITY EYES

A REVOLUTION IS BREWING. We're on the verge of being able to "see" black holes in action. For 400 years, astronomers have learned about the universe solely by using light and other forms of electromagnetic radiation. They measure properties of the "stuff" of the universe by the ways it emits and interacts with radiation. Then, in 2015, gravitational waves were detected for the first time.

Gravitational waves are ripples in space-time that travel at the speed of light. They offer a unique window into the intense gravity of black holes, neutron stars, and supernovae, and will allow astronomers to test general relativity in new ways. They reach us from vast distances and can be used to probe the universe just after the big bang. Seeing with gravity eyes promises to transform our understanding of black holes.

A New Way of Seeing the Universe

There have been two major revolutions in the way we see the universe. The first began in 1610, when Galileo took a newly invented device called a telescope and aimed it at the night sky. His best telescope had lenses a half inch across, and it gathered 100 times more light than the eye. Ever since Galileo's time astronomers have worked to improve his simple spyglass. A hundred years ago they began using mirrors instead of lenses to gather the light, since lenses sag when they're large and they don't bring all colors to a focus at the same location. In the modern era, astronomers have built optical telescopes 10 meters in diameter, either using single monolithic mirrors or mosaics of smaller hexagonal segments.[1] The gain in light-gathering power in four centuries since the time of Galileo is a factor of a million.

Meanwhile, an additional gain in depth came from improving the way light is detected. The eye is an inefficient, chemical detector. To give us the illusion of continuous motion it must transmit the information that falls on the retina to the brain 10 times a second. That means that it only gathers light, or "integrates," for a tenth of a second. Photography was invented in the mid-nineteenth century, and soon afterward astronomers were using it to take pictures of the night sky. Light is captured chemically in a process that's no more efficient than the eye, but long exposure times give much greater depth. A real leap forward came in the 1980s, when digital imaging was perfected. Charge-coupled devices, or CCDs, now have 80–90% efficiency in converting incoming photons into electrons and then into an electrical signal that can easily be digitized. CCDs are near-perfect detectors. The gain in detection efficiency over the eye is a factor of 100,000.

Combining these two factors means that the best telescopes see a remarkable factor of 100 billion times deeper than the eye. This is the difference between a Northern Hemisphere dweller seeing just one external galaxy, M31, and a large telescope seeing 100 billion. It's the difference between seeing stars a few hundred light years away and seeing light that has traveled for 13 billion years. CCDs have improved so much that the number of photons recorded with large telescopes in the past year exceeds the number of photons recorded by all the human eyes in history.

The second revolution in seeing the universe played out over the first half of the twentieth century. Ever since our early ancestors stared at the sky from the African savannah, astronomy has used a small sliver of the electromagnetic spectrum. From the bluest blue to the reddest red is only a factor of 2 in wavelength or frequency. The largest telescopes simply drill deeper in the same narrow slice of spectrum.

Technologies were developed to pry open the electromagnetic spectrum for astronomy. Viewing the universe in visible light is as limited as seeing in black and white compared to seeing in vivid color. Perhaps a better analogy comes from music: visible light is two adjacent keys on a piano, while the electromagnetic spectrum from radio waves to gamma rays is the full set of 88 keys. The first invisible waves to be used for astronomy were radio waves. At the end of the nineteenth century, Guglielmo Marconi showed that radio waves could be sent and detected over large distances, and, as we've seen, within thirty years Karl Jansky used a simple antenna to detect radio waves from the center of our galaxy. In the 1920s two astronomers at Mount Wilson Observatory used a device that converts a temperature difference into an electrical signal to detect infrared radiation from a number of bright stars, but infrared astronomy didn't take off until more sensitive detectors were perfected in the 1970s. Observations at invisibly short wavelengths were impossible until astronomers could avoid the radiation being absorbed by the Earth's atmosphere. The X-ray Sun was detected by a sounding rocket in 1949, and the archetypal black hole Cygnus X-1 was first spotted fifteen years later. X-ray astronomy advanced rapidly, with a series of satellites in the 1970s. Cosmic gamma rays were predicted years before they were detected by satellites in the 1990s.[2]

These capabilities give astronomers tools to detect radiation with wavelengths as long as 10 meters and as short as a thousandth of the size of a proton (frequencies from 10^8 to 10^{27} Hz). The extension in wavelength grasp from a factor of 2 to a factor of 10 billion billion attests to the power of technology to transform our view of the universe. Only a few sources can be detected at all wavelengths across the electromagnetic spectrum, and they're all active galaxies powered by supermassive black holes.[3]

Everything we learn about the universe involves telescopes gathering radiation. It's very easy to forget that we rely on indirect information. The universe is full of matter: dust grains, gas clouds, moons, planets, stars, and galaxies. We don't see this matter directly; we infer its properties by the way it interacts with electromagnetic radiation. Chemical elements are diagnosed by the particular spectral lines they emit or absorb. Dust grains reveal themselves by absorbing light and emitting infrared radiation. Moons and planets are seen in the reflected light of nearby stars. Stars are seen by the radiation they leak out as a byproduct of nuclear fusion. Galaxies are mapped using Doppler shifts of spectral lines from their gas and stars.

All of this is indirect, and it only relates to the 5% of the universe that's normal matter. The 95% that's dark matter and dark energy is still invisible to us because it doesn't interact with radiation. The astronomical objects are the actors, but the "stage" for this cosmic drama is also unseen. Astronomers trace the expansion of the universe using galaxies as markers of invisible space-time.

The detection of black holes is also indirect. The closest we get is information from the high-energy radiation in the surrounding corona that reflects off the inner part of the accretion disk; then the mass and spin of the black hole can be diagnosed through X-ray spectral lines.

Wouldn't it be nice to see the "stuff" of the universe without the intermediary of electromagnetic radiation? Wouldn't it be great to directly perceive the warping of space-time? We could, if only we had "gravity eyes" (Figure 52). The best analogy for what that might be like for a person is telepathy. A brain is a lump of living tissue weighing about three pounds. In more detail, it's an electrochemical network consisting of billions of neurons and trillions of connections between them. But this knowledge hasn't told us where we store memories, emotions, momentary thoughts, or our sense of self. Seeing the universe in terms

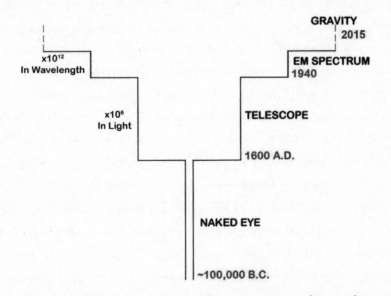

FIGURE 52. There have only been three revolutions in our way of seeing the universe. For most of human history, we were limited to naked-eye astronomy. In 1610, Galileo used the telescope as a way of gathering more light, and in the four centuries since telescopes have reached over 10 meters in diameter. In the early part of the twentieth century, a series of technological advances, involving new detectors and telescopes in space, opened up the electromagnetic spectrum for astronomy, from radio waves to gamma rays. In 2015, the detection of gravitational waves allowed us to see the universe with "gravity eyes" for the first time. *Chris Impey*

of gravity would be as profound as seeing someone else's thoughts and feelings as they experience them.[4]

Ripples in Space-Time

What are these ripples in space-time? Recall that in general relativity matter governs the curvature of space-time. Gravitational waves happen any time a mass changes its motion or its configuration.[5] Waves of distorted space radiate out from the source in the way that waves move away from a stone thrown into a pond. In the theory, the waves travel at the speed of light and they weaken with distance from the source. The space distortion is extremely subtle for most matter in motion. The strongest gravitational waves come from the most dramatic cosmic events: black holes orbiting each other and colliding, neutron stars orbiting each other and colliding, supernova explosions, and the violent birth of the universe itself.

Imagine perfectly flat space-time with a circular ring of particles lying in a plane. I think of it as the plane of my computer screen. The particles only serve the purpose of making invisible space-time visible. If a gravitational wave passes directly into or out of the screen, the ring of particles will follow the distortion in space-time, alternately squashing slightly in a vertical and then a horizontal direction, repeating the distortion periodically (Figure 53).[6] Like other waves, gravitational waves are described by their amplitude, frequency, wavelength, and speed. The amplitude is the fraction by which the ring of particles distorts as the wave passes. The frequency is the number of times the ring of particles stretches or squeezes per second. The wavelength is the distance along the wave between points of maximum stretching or squeezing. These waves travel at light speed through the cosmos, flexing physical objects but also passing through them as if they weren't there.[7]

In the analogy we imagine a circle flattening and stretching into an ellipse. But that far overstates the actual distortion for a typical gravitational wave. The imaginary ring of particles deviates from a circle by 10^{-21}, or one part in 1,000 billion billion! To detect space-time shimmering by such a tiny amount sounds like an impossible experiment.

Indeed, the person whose theory predicts gravitational waves at first did not believe they were real. We've seen that Einstein didn't believe that black holes existed, and he underplayed the importance of gravitational lensing. In 1916, following a suggestion by his colleague Henri Poincaré, he made an analogy with electromagnetism. When an electric charge is moved back and forth the oscillating disturbance creates an electromagnetic wave such as light. Einstein knew that matter curves

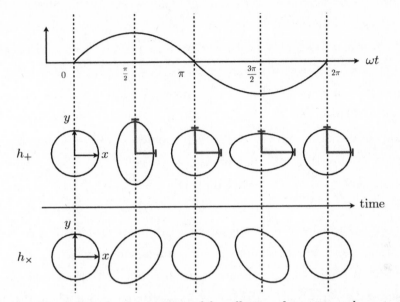

FIGURE 53. Gravitational waves are sinusoidal oscillations of space-time, shown in the top panel as one complete cycle of amplitude variation. Two independent polarizations of gravitational waves, shown in the lower two panels, are displaced by 45 degrees, rather than the 90 degrees of electromagnetic waves. The wave moves perpendicular to the page in this example. The middle panel shows the arms of an interferometer, added to suggest how such waves might be detected. Distortion in these diagrams is highly exaggerated; in practice, deviations from circles would be imperceptible. *S. Mirshekari/ Washington University, St. Louis, used with permission*

space, so it seemed logical that matter in motion should create an oscillating disturbance of space.

Einstein had terrible problems making this idea work. The analogy is flawed because electric charges can be positive and negative while in gravity there is no such thing as negative mass. Einstein struggled mightily with coordinate systems and approximations to do the necessary calculations. He derived three types of wave, then was mortified when Arthur Eddington showed that two were mathematical artifacts that could travel at any speed. Eddington joked deadpan that they might even "propagate at the speed of thought."[8]

By 1936 Einstein had made up his mind. Every time he tried to write a formula for plane waves, like our computer screen analogy, he encountered a singularity where the equations blew up and the quantities became infinite. He wrote a paper with Nathan Rosen, his student at Princeton, titled "Are There Any Gravitational Waves?" to which the answer in the paper was an emphatic "No!" He submitted it to the prestigious journal *Physical Review* and was stunned when an anonymous referee rejected it and pointed out several errors. Einstein had never been subjected to peer review before, and in Germany his papers had always been published automatically. He wrote a testy letter to the editor: "We (Mr. Rosen and I) had sent you our manuscript for publication and had not authorized you to show it to specialists before it is printed. I see no reason to address the— in any case erroneous—comments of your anonymous expert."[9]

But Einstein was wrong, and another young colleague pointed out his mistake—ironically, a day before he was due to give a talk at Princeton titled "The Nonexistence of Gravitational Waves." After Einstein and Rosen published their corrected paper in another journal, physicists remained divided.[10] Many thought that gravitational waves were a mathematical construct with no physical meaning. But after all his early misgivings, Einstein became convinced they were real. The success of his theory gradually persuaded him to trust its predictions.

An Eccentric Millionaire and a Solitary Engineer

Space-time ripples seemed so difficult to detect that physicists ignored them. They were buried in the drawer of physics esoterica for twenty years after Einstein and Rosen's paper was published, until they caught the attention of an eccentric American millionaire named Roger W. Babson. If you never thought physics could make you rich, pay close attention to this story.

Babson's interest in gravity started with a family tragedy. His older sister drowned while he was still an infant, and he later said it was because she couldn't fight gravity. In his career, he applied a version of Newton's laws to the stock market. "What goes up must come down," he said, and also "to every action there is a reaction."[11] He foresaw the 1929 Wall Street crash, and he generally managed to buy cheap shares on their way up and sell them before they headed back down.[12] Babson said he owed gravity a debt for helping to make him a millionaire.

Babson started the Gravity Research Foundation in 1949, and sponsored a high-profile essay contest on ways to counteract or nullify gravity. Needless to say, the contest was won by some less than rigorous essays. The Foundation's promotional material talked about control of gravity in the context of Jesus walking on water.[13] Reputable physicists avoided it and science popularizer Martin Gardner called the foundation "perhaps the most useless project of the twentieth century."[14]

In an attempt to regain credibility with the physics community, Babson spun off an institute whose sole purpose was to fund pure research into gravity. He asked the Princeton physicist John Wheeler, who invented the term "black hole," to persuade his colleague Bryce DeWitt to head the new institute. DeWitt organized a landmark conference on gravity and general relativity at the University of North Carolina in early 1957.

The conference energized a young generation of gravity theorists.[15] The discussion on gravitational waves centered on whether or not they carry energy. Richard Feynman's "sticky bead" argument convinced most of the audience. He asked them to imagine two separate rings of beads that fit snugly on a metal bar. When a gravity wave passes through the bar, its force causes the rings to slide back and forth slightly. Friction between the rings and the bar means the bar will heat up. Thus, energy is transmitted from the wave to the bar. A young engineer named Joseph Weber was in the audience, and he paid careful attention to this talk.

Weber was born into a poor Lithuanian immigrant family, his name Anglicized to ease assimilation. He dropped out of college to save his parents money and joined the Navy, rising to the rank of lieutenant com-

mander. During World War II he headed electronic countermeasures
work for the Navy. After the war, he joined the engineering faculty of
the University of Maryland. Weber's scientific life was a series of near
misses. George Gamow could have given him the PhD project of detecting microwaves from the big bang, but he didn't, and the Nobel Prize
for the later, accidental, discovery was won by Arno Penzias and Robert
Wilson. In 1951, Weber presented the first paper on the idea of masers
and lasers, but it was Charles Townes who, after reading his paper, pioneered these technological innovations. But Weber's most painful miss
was gravitational waves.[16]

Inspired by the conference in Chapel Hill, Weber wondered how he
might detect gravitational waves. He came up with the idea of a metal
cylinder suspended on wires and placed in a vacuum chamber to isolate
it from the environment. His cylinder was 1.5 meters long and two-thirds
of a meter in diameter, and weighed 3 metric tons. It was surrounded by
piezoelectric sensors to convert mechanical vibrations into electrical signals.[17] If a gravitational wave passed through the cylinder, Weber hoped,
it would ring just like a bell hit with a hammer (Figure 54).

Weber set up one of his "bars" in a room at the University of Maryland and an identical one 600 miles away at Argonne National Laboratory outside Chicago. Their data link was a high-speed phone line.
The reason to have two identical detectors was to eliminate local noise
caused by thunderstorms, mild earthquakes, cosmic ray showers, power
glitches, and anything else that might jostle the cylinder. If a signal
wasn't recorded simultaneously at both locations, it would be discarded
as spurious. Apart from local events, the constant noise in Weber's
experiment was the thermal motion of atoms in the aluminum cylinder. This unavoidable agitation caused the cylinder to vary erratically in
length by roughly 10^{-16} meters, or less than the size of a proton.

Weber thought he had hit the mother lode when he saw signals that
were well above the level of this thermal noise. In 1969, he published
the detection of gravitational waves and announced it at one of the major

FIGURE 54. Joseph Weber and his pioneering gravitational wave detector, set up in a physics lab at the University of Maryland. An identical detector was set up 600 miles away; a real signal would register with both detectors. Like Grote Reber before him, Weber was for a time the only person doing experiments in this new field of astrophysics. He announced the detection of gravitational waves in 1969, but nobody could replicate his results so his reputation suffered. V. *Trimble, used with permission*

Gravitational and Relativity meetings. A year later, he claimed that many of the gravitational waves originated from the center of the Milky Way galaxy.[18] Physicists were all surprised, and many were stunned. But most were delighted that a central prediction of general relativity had been confirmed. Weber was celebrated. His picture appeared on magazine covers. He was famous.

Then it all began to unravel. Weber's signal from the center of the Milky Way implied that 1,000 solar masses per year were being converted into gravitational wave energy. The young theorist Martin Rees calculated that such a loss of mass would cause the galaxy to become "unglued" and fly apart. Other experimenters tried to replicate Weber's results. Weber bars sprung up around the United States and in Germany, Italy, Russia, and Japan. Ron Drever, whom we'll meet later, set up several in Glasgow (a place not exactly short of bars). There's even a Weber bar on the Moon, left there in 1972 by Apollo astronauts. By the mid-seventies, several groups had improved on Weber's original design and sensitivity, often cooling their detectors to reduce thermal noise.

Nobody detected anything. Other physicists questioned Weber's experimental technique. He seemed to have miscalculated the statistics of coincident events in his widely separated detectors. Damningly, he claimed a peak in his data every twenty-four hours, when the center of the galaxy passed overhead. It was quickly pointed out that gravitational waves should pass through the Earth like a knife through butter, so he should have seen a peak every twelve hours. In 1974, at the seventh major Gravitation and Relativity conference, senior IBM physicist Richard Garwin denounced Weber and his data.

Soon, the rest of the physics community agreed. Weber had been guilty of poor experimental technique and worse: bias in his presentation of data. Despite this, he never wavered in his belief that he had seen ripples in space-time. By the end of his career he was a largely embittered and solitary figure.[19]

Yet Weber's work spurred innovation, as other physicists were motivated to detect this signature prediction of general relativity. John Wheeler wrote:

Following our work together in Leiden, he embraced gravitational waves with religious fervor and has pursued them for the rest of his professional career. I sometimes ask myself whether I imbued in Weber too great an enthusiasm for such a monumentally difficult task. Whether, in the end, he is the first to detect gravitational waves or whether someone else, or some other group does it, hardly matters. In fact, he will deserve the credit for leading the way. No one else had the courage to look for gravitational waves until Weber showed that it was within the realm of the possible.[20]

Despite the discouraging results of the Weber experiments, there was a glimmer of hope. In 1974, Joe Taylor and Russell Hulse were using the Arecibo 305-meter radio telescope for pulsar observations. They found a pulsar spinning 17 times a second, but noticed a systematic variation in the arrival of the pulses. The variations had a period of eight hours, which suggested a binary system. Additional observations showed that PSR 1913+16 was a pair of neutron stars in a tight orbit not much larger than the size of the Sun. Taylor and Hulse realized that general relativity predicted orbital decay in the binary system: the orbital period should decrease by 77 microseconds per year as energy is carried away by gravitational waves. Pulsars are exquisite clocks, so the tiny period shift was observable (Figure 55). The orbital decay detected precisely matched the prediction of general relativity.[21] This was strong, although indirect, evidence for gravitational waves.[22] Taylor and Hulse were awarded the Nobel Prize in Physics in 1993 for making this sublime observation.

The binary pulsar pointed the way forward. Another dozen systems were discovered. Astronomers realized that binary black holes should exist as well, with stronger gravity and therefore more powerful gravitational waves. With a sufficiently sensitive detector, perhaps these waves could be detected directly.

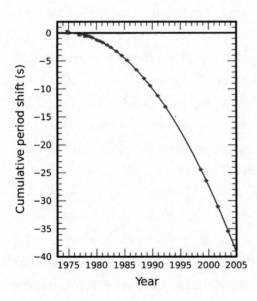

FIGURE 55. Orbital decay of the binary pulsar system PSR 1913+16, as observed at the 305-meter Arecibo radio dish by Russell Hulse and Joe Taylor. The system loses energy by emitting gravitational wave radiation, and the data points perfectly fit the theoretical prediction of general relativity. These observations were a strong affirmation of the validity of general relativity and indirect confirmation that gravitational waves exist. *Inductiveload*

When Black Holes Collide

This is the story of how two black holes formed, and how in the blink of an eye their collision unleashed a torrent of gravitational waves containing 10 times more power than the light of all the stars in the universe. It's also the story of the birth of a new field of astronomy.

It's 11 billion years ago and the universe is a cozy place, 3 times smaller and 30 times denser than it is now. This is the "construction phase" of the universe, when galaxies are small and dense, merging with one another and actively forming stars. In one small, unremarkable galaxy, two massive stars form close to each other in a chaotic region of gas and dust. They're 60 and 100 times the mass of the Sun—as hefty as stars can get. Within a few million years—a cosmic snap of the fingers—both stars consume their nuclear fuel. The more massive star lives faster and dies first, but as it becomes old and bloated its smaller companion steals gas from it, overtakes it in mass, and becomes a black hole first. The black hole sucks gas from its companion, cloaking the pair in a shroud of gas that's churned up by the orbital motion. The gas also sucks energy from the orbit, bringing the two stars as close as Mercury is to the Sun. Then the second star dies and becomes a black hole.

At the end of this phase of vampirism, two black holes remain. Each hides 30 times the Sun's mass behind the impenetrable veil of an event horizon 150 miles in diameter. They orbit each other warily, locked in a gravitational embrace.[23]

For 10 billion years, nothing happens. The pair orbit in silence and in darkness, eking out a tiny dribble of gravitational waves that brings them imperceptibly closer. Beyond, the universe gets larger, older, and colder. The cosmic expansion rate transitions from deceleration to acceleration as the baton passes from dark matter to dark energy. Star formation peaks and declines, and, no doubt, on surfaces of numerous Earth-like planets, alien civilizations rise and fall as well. Meanwhile, on the planet that we call home, 3 billion years after life started it is still exclusively microbial.

Then comes a crescendo of activity. As the black holes approach each other the gravity gets stronger and more gravitational waves are emitted, shrinking the orbit and accelerating the process. The last phases take just two-tenths of a second. The black holes increase their orbital speed and enter a death spiral. Space-time is churning like a pot of water brought to a hard boil. Gravitational waves are created with a frequency that matches the orbital period. It rises quickly from 35 Hz to 350 Hz. To approximate this in sound, you'd have to whip your hand up a piano keyboard from the lowest A to middle C in a fraction of a second. Think of a familiar orbit, like the Moon's around the Earth. That takes a month at a separation of a quarter of a million miles. At the end of the death spiral these two black holes, each 10 million times more massive than the Earth, are about 100 miles apart and hurtling around each other 300 times in a second, or at half the speed of light. This isn't an orbit—it's insanity.

Then the event horizons "kiss" and the black holes merge. The equations for this can't be solved; even supercomputers strain to calculate what happens. The last phase is called ring-down, where the merged object oscillates like a huge glob of dark Jell-O before settling down into a single black hole twice the mass and twice the size of the individual black holes that collided (Figure 56). In the gravity math, 5% of the mass is converted into gravitational waves. A million Earth masses turns into the energy of space-time ripples and escapes the entombment in the black hole. (By contrast, the Sun in one second converts a thousand-trillionth of the mass of the Earth into radiant energy.) A pulse of gravitational waves flees the scene at the speed of light, traveling in all directions like waves in a three-dimensional pond. The biggest explosion ever recorded in the universe occurs in silence and total darkness.

The ripples sweep through the voids of intergalactic space, weak-

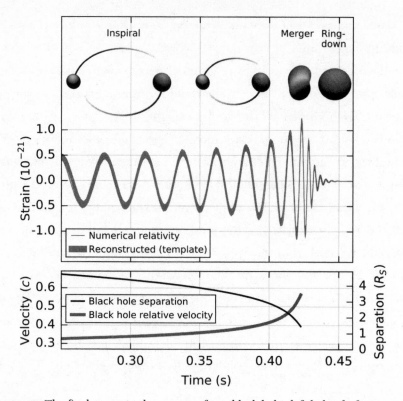

FIGURE 56. The final stages in the merger of two black holes left behind after massive stars have evolved and died as supernovae. After a slow approach lasting millions of years, the black holes spiral toward each other, merge, and reverberate before settling down. This entire sequence takes less than 0.2 seconds. The middle panel shows the gravitational wave signal predicted by numerical relativity, and the lower panel shows that the two black holes merge at a speed more than half the speed of light. *LIGO Scientific Collaboration/Institute of Physics*

ening as they move away from their source. They pass through millions of galaxies, perhaps without being noticed. Meanwhile, on Earth, life moves from the oceans onto the land, the dinosaurs emerge and are extinguished by a global catastrophe, and a branch of the primate line develops large brains. As the gravitational waves wash across our neighboring galaxies, the Magellanic Clouds, our ancestors learn how to harness fire. As they enter the Milky Way, humans leave Africa for the first time. The waves pass near the bright star Beta Volantis as Albert

Einstein is publishing his new theory of gravity. They pass close to a neighboring dwarf star 82 Eridani as a huge scientific instrument begins construction at widely separated locations in the United States. The instrument is taken offline for five years for upgrades, and it's getting ready to take science data for the first time as the waves sweep through the Solar System and bear down on the Earth.

Marco Drago sits bolt upright. The thirty-two-year-old postdoc from Italy is nursing a cappuccino at his computer monitor in the Albert Einstein Institute in Germany when he sees a small squiggle on the monitor. Software initially flags the event as a glitch, but after automated cross-checks the flag is removed. Marco realizes the universe is speaking, so he composes an email with the subject line "Very interesting event." He's at the helm of the most precise machine ever built.

The Most Precise Machine Ever Built

How many physicists does it take to measure a shift of a ten-thousandth of the width of a proton? Answer: over a thousand. Marco Drago is one of a small army of scientists working at dozens of universities and research institutions around the world on the most sensitive scientific instrument ever conceived. The story of how the Laser Interferometer Gravitational-Wave Observatory, or LIGO, got built is almost as unlikely as the detection of ripples in space-time.

When we left the story of gravitational wave detection, the field was in disarray. Nobody could reproduce Weber's results and his scientific reputation was in tatters. Unfair as it might seem, the taint was widespread. Gravitational wave hunters were charlatans or fools, maybe both.

But one group of researchers was motivated by their inability to reproduce Weber's results. It was a challenge to them as experimentalists to do better. They were buoyed by the pulsar spin-down observations of Taylor and Hulse, which was evidence that gravitational waves

existed. One of these men (for it was and is a male-dominated field) was
MIT physicist Rainer Weiss. As a child, Weiss fled Nazi rule in Germany
with his family. He grew up in New York in a situation of benign neglect,
and threw himself into his passions of classical music and electronics.
He abandoned his classes at MIT, and had to start from the bottom as
a technician in a physics lab. He returned to MIT but struggled to get
tenure. And he grew frustrated trying to explain Weber's results to his
students. "I couldn't for the life of me understand what Weber was up
to," he said. "I didn't think it was right. So I decided to go at it myself."[24]

Weiss worked for a whole summer in isolation in a basement on an
idea that grew out of discussions with his MIT students,[25] and came up
with a detector that was an interferometer instead of a single bar. Imagine two metal bars at right angles, forming an L-shape. If a gravitational
wave arrives from above, the way it squeezes and stretches space means
it makes one bar very slightly shorter and the other bar very slightly longer. An instant later the opposite happens, and the pattern repeats as
long as the wave is active. Instead of trying to detect a single bar ringing
like a bell, Weiss would have to detect two bars alternately flexing.

Weber's experiment had been thousands of times too insensitive to
detect its quarry; Weiss knew he had to make dramatic improvements.
His clever idea was to use light as a ruler. His "bars" would be long
metal tubes with the air sucked out, since light travels at constant speed
in a vacuum. A laser at the crook of the L sends light of one wavelength
through a beam-splitter, so half goes down one arm and the other half
goes at right angles down the other. The light bounces off a mirror at
the end of each arm, comes back to the crook of the L, and recombines
at a detector. Normally, light waves return down each arm in perfect
synchrony, peaks and troughs in lockstep. But when a gravitational wave
passes through the instrument, one beam travels a slightly shorter distance than the other, so the peaks and troughs don't line up and the
light intensity is reduced (Figure 57).

It sounds simple enough. The challenge is the exquisite precision of

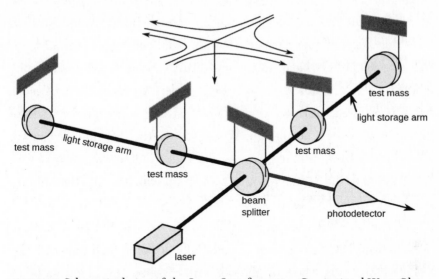

FIGURE 57. Schematic design of the Laser Interferometer Gravitational-Wave Observatory (LIGO). The gravitational wave is depicted as arriving from directly overhead. Light passes through a beam splitter and along arms 4 kilometers long, returning to be recombined at the photodetector. The test masses along each arm register the arrival of a gravitational wave as very slight changes in the length of the arms, which is recorded as an interference pattern by the photodetector. *California Institute of Technology/MIT/ LIGO Laboratory*

the measurement. Not only is the amplitude of a space-time ripple very small, its wavelength is very long. The typical frequency of the gravitational waves from a black hole collision is 100 Hertz, which means that 100 ripples pass by every second. But the typical wavelength is 3,000 kilometers. The optimum length of the arms of such a device is a quarter of the wavelength, since shift by a quarter wave in either direction is the difference between enhancing the signal and canceling it. Weiss knew he couldn't make a 750-kilometer vacuum tube, so he imagined bouncing the light back and forth many times in a shorter tube. Weiss wrote up his concept in an MIT technical report in 1972. It may be the most influential paper never to be published in a scientific journal.[26]

The early road was hard. Weiss started work on a prototype interferometer with 5-foot-long arms. Even though it was hundreds of times smaller and less expensive than any viable tool to detect gravitational waves, he still had trouble scraping up enough funds. Administrators were dubious, and one particularly influential colleague, Philip Morrison, was deeply skeptical. In the early 1970s, there wasn't even strong evidence that Cygnus X-1 was a black hole. Morrison thought black holes didn't exist, and since they were the strongest potential sources of gravitational waves, he thought Weiss was wasting his time. Weiss got some money from military sources, but that funding was cut off by an amendment to the Military Authorization Act which barred the military from supporting civilian projects.

One summer day in 1975, Rainer Weiss went to Dulles Airport in Washington, D.C., to pick up famed Caltech theoretical physicist Kip Thorne—the man who won the bet with Stephen Hawking that black holes existed. Science works best when theory and observation work in concert. Theoretical predictions drive better observations and observations drive deeper physical understanding. The genesis of the LIGO project occurred that steamy afternoon in Washington, when the experimentalist Weiss encountered one of the great theorists of our time in Thorne.

Weiss had invited Thorne to a meeting at NASA headquarters on doing cosmology and relativity research in space. "I picked Kip up at the airport on a hot summer night when Washington was filled with tourists. He didn't have a hotel reservation so we shared a room for the night," recalled Weiss. "We made a huge map on a piece of paper of all the different areas in gravity. Where was there a future? Or what was the future, or the thing to do?"[27] They were so consumed by their conversation that neither of them slept.

Thorne hadn't read Weiss's technical paper on the interferometer concept. "If I had," he said later, "I had certainly not understood it." In fact, his magisterial book *Gravitation* includes a student exercise designed to show the impracticality of detecting gravitational waves with lasers. "I turned around on that pretty quickly," admitted Thorne.[28]

He returned to Caltech fired up to build an interferometer. But first he needed to recruit an experimental physicist. Weiss suggested Ron Drever of the University of Glasgow. Drever had done fundamental experiments on the smoothness of space and the mass of neutrinos, he'd built and operated a Weber bar, and he'd constructed an interferometer with 10-meter arms, 6 times bigger than the modest instrument built by Weiss at MIT. Thorne got Drever a half-time faculty position at Caltech, and by 1983 he'd built an interferometer with 40-meter arms there, using ingenious methods to increase the laser power and improve the isolation from seismic noise.

Money started to flow and the competition started to heat up. Weiss received a small grant from the National Science Foundation in 1975 to start his interferometer work. In 1979, the Caltech group led by Thorne and Drever got a significant grant, while the MIT group led by Weiss got a smaller amount of money. Caltech and MIT are intense science rivals;[29] the Caltech group, with their 40-meter interferometer, were clearly in the lead. Weiss must have regretted recommending Drever to Thorne. Both groups were dreaming of a full-scale, kilometer-sized interferometer, but it was Weiss who seized momentum with a visit to the National Science Foundation. He pitched the concept of an interferometer on two sites with a price tag of $100 million. The design study that resulted was called the "Blue Book," and it's the de facto bible for detecting ripples in space-time.[30]

Weiss and Drever were both intensely competitive. Thorne found himself in the role of intermediary and peacemaker. As the NSF made it clear that the two groups would not be funded separately, they found themselves in a shotgun marriage. Progress came in fits and starts. Continual delays over technical issues led the NSF to cancel funding.[31] By the mid-1990s LIGO was back on track, now led by Caltech high-energy physicist Barry Barish. Science is littered with projects where accomplished scientists failed because they lacked interpersonal and management skills, but Barish proved to be an adept manager.

From the start, they had planned on two identical interferometers with 4-kilometer arms located on opposite sides of the U.S., in geologically quiet sites. One was near a mothballed nuclear reactor in scrub desert outside Hanford, Washington, and the other was in swampland outside Baton Rouge, Louisiana. The first stage, called initial LIGO or iLIGO, had a goal of technology development; actual detection was very unlikely. The second stage, advanced LIGO or aLIGO, aimed to be sensitive enough to detect the gravitational waves that theory predicted (Figure 58). Barish wanted a facility and an infrastructure where all the major components—vacuum systems, optics, detectors, and suspension systems—could be continuously improved.

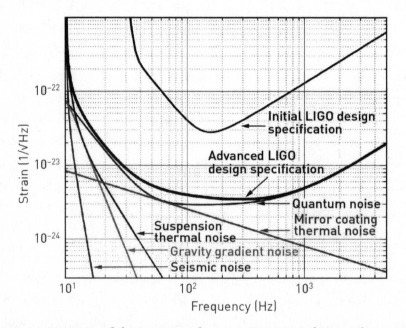

FIGURE 58. Sensitivity of the Laser Interferometer Gravitational-Wave Observatory (LIGO). This is characterized in terms of "strain," which approximates to the fractional change in length of the test mass when a gravitational wave passes directly through it. At low frequencies the sensitivity is bounded strongly by geological noise and effects of gravity on the mirrors. At high frequencies the sensitivity is bounded by quantum noise in the detectors. The gain in sensitivity going from initial LIGO to advanced LIGO is at least a factor of 10. *LIGO Scientific Collaboration*

The far greater sensitivity of advanced LIGO required upgrades to almost every aspect of the experiment. The laser became more powerful, to reduce the principal source of high-frequency noise. The test masses at either end of each arm were made heavier; each test mass was a 40-kilogram cylinder of silica with a mirror attached, designed to detect tiny changes in the length of the arm. A four-stage pendulum was used for suspension, and the isolation and noise cancelation was improved by an order of magnitude. LIGO has the largest and best vacuum system ever built, requiring 30 miles of welds with no leaks. The pipes are so long that they're a meter off the ground at each end as the Earth curves under them. The most precise concrete pouring and leveling ever achieved was needed to counteract this curvature and keep the pipes flat and level. The vacuum is a trillionth of the density of air at sea level. The detectors are so sensitive they can feel when a truck applies its brakes three miles away and hear lightning storms at a distance of 50 miles. Even more impressive, they can see the motion of individual atoms in their mirrors.

The experiment is a technical tour de force. Initial LIGO ran from 2002 to 2010 and, as expected, it didn't detect gravitational waves. The upgrade to advanced LIGO took five years and the work of 500 people. Advanced LIGO worked in an engineering mode for six months and hit paydirt four days before it was due to start taking science data.

Which brings us back to Marco Drago and the morning of September 14, 2015. The soft-spoken postdoc plays classical piano and has written two fantasy novels in his time away from physics. When he saw the squiggle on his screen he was immediately suspicious. It had the classic pattern of a black hole merger, a brief crescendo that researchers call a "chirp," like the bird song of the universe. After over a billion years of travel across space, the wave passed through the Earth at the speed of light, jostling the detector at Livingston in Washington seven milliseconds before it jostled the detector across the country at Hanford in Louisiana (Figure 59). Drago was skeptical because the signal looked

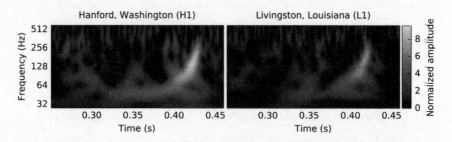

FIGURE 59. GW150914, the first gravitational event ever detected. The classic "chirp" pattern of merging black holes is seen in a graph of gravitational wave amplitude vs. frequency. The signal arrived at the LIGO Hanford detector 7 milliseconds before it arrived at the LIGO Livingston detector—consistent with the time taken for gravitational waves to travel between the two sites. *LIGO Scientific Collaboration/Institute of Physics*

too strong, too perfect: "No one was expecting something so huge, so I was assuming that it was an injection."[32] LIGO overseers kept the team on its toes by injecting false signals into the data stream; they're called "blind injections." In 2010, a blind injection led to great excitement and a paper being written, and it was only when the paper was about to be submitted that the team was told that the signals were fake.

Drago took pains to be sure. He called all the other sites and talked to a group leader to make sure nobody had injected a signal into the system. He even worried that someone might have hacked the system as a prank. After a dozen automatic and manual checks, there was no doubt that the universe was responsible for the signal. It stood out against the noise like a burst of laughter in a room full of chattering people. Gravity had spoken.

Meet the Maestro of Gravity

The world's preeminent gravity theorist originally wanted to be a snowplow driver. Kip Thorne was at home with snowstorms and mountains as a child. "Growing up in the Rocky Mountains, that's the most glorious job you can imagine. But then my mother took me to a lecture about the Solar System and I was hooked."[33] He was raised Mormon in a conservative area of Utah, but is now an atheist. His parents were both academics so they encouraged his curiosity.

Thorne's career moved swiftly. After earning degrees at Caltech and Princeton, he returned to Caltech, where he was one of the youngest people ever to be made a full professor. He'd left Utah as a gaunt and geeky Mormon, his shyness hidden behind a messianic beard. By age thirty he was a world expert on gravitational astrophysics who favored jeans, a black leather jacket, and a hipster goatee.

Thorne did his PhD thesis under John Wheeler at Princeton.[34] Wheeler posed an intriguing question: would a cylindrical bundle of magnetic field lines implode under its own gravitational force? Magnetic field lines repel one another, and after difficult calculations Thorne proved it was impossible for a cylindrical magnetic field to implode. This led to another question. Why then are spherical stars, which are also threaded by magnetic field lines, able to implode and become black holes? Thorne figured out that gravity can only overcome interior pressure when it acts in all directions. Imagine a hoop that can be spun to trace a sphere. Any object of mass M around which a hoop with circumference $4\pi GM/c^2$ can be spun must be a black hole (where G is the gravitational constant and c is the speed of light). This was called the Hoop Conjecture, and it made Thorne a superstar when he was barely out of graduate school.

By the time he was in his mid-thirties, Thorne had coauthored his landmark textbook, *Gravitation*, and embarked on his series of bets with Stephen Hawking. As a cofounder of LIGO, Thorne was deeply invested in discovering gravitational waves. He knew that the merger of two black holes would give the strongest signature of gravitational waves. But there was a problem. The only way to calculate the strongest part of the signal, just before the merger, was by using computation, since, as is the case for many situations in general relativity, the equations could not be solved exactly. However, supercomputer simulations at the time were woefully inadequate.

> It was very much on our minds that those computer simulations needed to be in hand by the time we might begin to see gravitational waves with LIGO. But in the 1990s, there were huge problems in the field. These great computational scientists could collide two black holes head-on, but when they tried to have the black holes go in orbit around each other, as should happen in nature, they couldn't even get them to go around once before the computers crashed. By 2001, I got alarmed, because I expected that Advanced LIGO would be operational in the early 2010s, a decade into the future. It was not at all clear that the simulations would be in hand by then.[35]

So he withdrew from the day-to-day operations of the project to start a numerical relativity group at Caltech and Cornell.

Thorne has a light touch as a popularizer of science. He's able to explain esoteric ideas in everyday language.[36] As the public face of LIGO, he was able to persuade career politicians with no science background to fork over nearly a billion dollars to build two huge machines to detect hypothetical, invisible waves so weak they can only shift atoms by a tiny fraction of their size.

His proximity to Hollywood means Thorne has been roped into projects where gravity is a major actor. In the early 1980s, Carl Sagan set him up on a blind date with producer Linda Obst and tapped his expertise to map out the wormhole travel scene in *Contact*. Obst then sought him out when she was developing *Interstellar* with director Christopher Nolan. The movie uses a gigantic spinning black hole called Gargantua to slow down time, and Thorne worked with the animators to ensure that the visuals were scientifically accurate. Some frames took 100 hours to render and the data for the film pushed

FIGURE 60. Artist's representation of the supermassive black hole called Gargantua in the 2014 film *Interstellar*. The true distortions of the accretion disk seen from this close would be even greater than in the image. In the film, the black hole holds the key for an astronaut to travel through time and space and save the people of Earth. Miller's planet is near the top left; it's an ocean world so close to the black hole that time slows dramatically. *O. James/Institute of Physics*

close to a million gigabytes. Thorne even made a scientific discovery from the simulations that will lead to several papers.[37] From his perspective the images are beautiful, but he thinks they're also beautiful because they're true (Figure 60).

Viewing the Universe with Gravity Eyes

Almost every day, a space-time ripple passes through your body due to a binary black hole cataclysm somewhere in the universe. It might approach from above, or the side, or from below your feet. You go about your business unaware of the intrusion. As the wave sweeps through, you get a tiny bit taller and thinner for an instant, then a tiny bit shorter and fatter, and the pattern repeats. After a few tenths of a second, you're normal again.

It brings to mind the reaction of the novelist and poet John Updike to a different ghostly messenger of the cosmos, the neutrino:

The earth is just a silly ball
To them, through which they simply pass,
Like dustmaids down a drafty hall
Or photons through a sheet of glass.
They snub the most exquisite gas,
Ignore the most substantial wall,
Cold-shoulder steel and sounding brass,
Insult the stallion in his stall,
And, scorning barriers of class,
Infiltrate you and me! Like tall
And painless guillotines, they fall
Down through our heads into the grass.[38]

There are three types of gravitational waves.[39] One is stochastic, a word that describes any physical process that has a random quality. This is the hardest type to detect, since the signal competes with random noise from electronics at high frequency and with geological activity at low frequency. The most exciting form of stochastic signal would be from the big bang, as we'll see shortly. The second is periodic, referring to gravitational waves whose frequency is nearly constant for a long time. The most common sources of periodic signals are neutron stars and black holes in orbit around each other. Since these binaries have wide separations, the signals are weak. The third is impulsive, meaning gravitational waves that come in a short burst. Bursts come from the formation of a black hole in a supernova explosion, and from the merger of neutron stars or black holes. They're expected to be the strongest sources of gravitational waves, and they have a distinctive fingerprint so they're also the easiest to distinguish from noise.

Think of black holes colliding as a gravity bell ringing. Just as a large bell makes a lower-frequency sound than a small bell, large masses colliding emit lower-frequency gravitational waves than small masses colliding. Neutron stars "chirp" in a crescendo up to 1600 Hz, minimum-mass black holes rise up to 700 Hz, and the hefty black holes detected in the first LIGO event started at 100 Hz and ascended to about 350 Hz. There are roughly 3 times as many neutron stars as black holes, so in order of decreasing number of events but increasing signal power, we expect to see: mergers of two neutron stars, mergers of a neutron star and a black hole, and mergers of two black holes. LIGO was designed to have maximum sensitivity in the frequency range 100–200 Hz, where merging black holes give their strongest signals. That's a sweet spot for detection. At 1,000 Hz, the sensitivity is 2 times worse because noise in the electronics rises, and at 20 Hz the sensitivity is 10 times worse as the geological rumbling of the Earth increases.

What information can we get from a space-time ripple? Let's use an analogy of water ripples. Imagine you're a cork on a large pond on a breezy day. The wind is ruffling the surface of the water and the random pattern of waves that makes you bob up and down is a good approximation of the background noise in a gravitational wave experiment. If someone starts dropping stones into the pond every second for a few seconds, you'll feel an additional bobbing motion that's periodic. That's the chirp from two black holes coming together. The size of the motion depends on the size of the stone, and on the distance from the place where the stone is being dropped, since ripples get weaker as they travel outward. As a cork, you have no eyes and ears so all you feel is the motion; you have no idea where the waves are coming from. But if you could talk to a second cork nearby, you'd have more information. The ripples travel in concentric circles, so timing receipt of the two signals gives you the direction of the source by triangulation. That's the way your ears work to figure out the direction a sound is coming from.

From a detection with LIGO, physicists can learn several impor-tant pieces of information.[40] The pattern of changing frequency gives the masses of the two black holes by comparison with a simulation. The merging phase is used to measure the spin of the post-merger black hole. Position of the event on the sky is measured using the time delay between the signal arriving at the two detectors (and the fact that a similar signal is seen at both locations helps to rule out noise or a spurious source for the signal). With only two sites, the sky position isn't very tightly constrained; it could be anywhere in a broad swath. However, LIGO's success has energized the international community. Europe has just commissioned an interferometer in Italy (Virgo) and one in Germany is not far behind (GEO600). An interferometer in Japan will go online in 2019, and another one is planned for India in the early 2020s. Detection at three or more sites might allow the sources of the gravitational waves to be pinpointed to a particular astronomical source and so be observed across the electromagnetic spectrum.[41]

The distance of the source is estimated by the strength of the signal. The waves move away from the black hole in three dimensions and are diluted as they spread through space. Gravitational waves have one great advantage over electromagnetic waves: their amplitude is inversely proportional to distance. If black holes are 10 times farther away, the signal is 10 times weaker. But astronomers can't measure the amplitude of an electromagnetic wave; they measure the intensity, which is the square of the amplitude, and if a star is 10 times farther away, the light intensity is 100 times weaker. That's why LIGO, reading gravitational waves, has enormous reach and is able to detect cataclysms billions of light years away.

What if LIGO's detection was a fluke? You can't do statistics with one event. Was the universe going to reveal its secrets in a single short song? Physicists were elated, but they were also anxious. They took solace in Einstein's words from a brief time in 1921 when it appeared that general relativity had been disproven by experiment: "God is subtle but he is not malicious."

There was excitement tinged with relief when the LIGO team announced the detection of a second event on December 26, 2015. The signal was weaker because the source was slightly farther away, at a distance of 1.5 billion light years, and because the black holes were smaller, 9 and 14 times the mass of the Sun as opposed to 29 and 36 times with the first event. An intervening event on October 12, 2015, was placed in the status of candidate rather than counted as a secure detection. It was weak because the black holes involved, 13 and 23 times the mass of the Sun, had merged not long after life began on the Earth, at the prodigious distance of 3.3 billion light years.[42] In 2017, LIGO made three more detections (Figure 61). With five firm detections and a sixth one on the bubble, a thousand scientists were jubilant. LIGO is a roaring success. It is the dawn of the age of gravitational wave astronomy.

In August 2017, LIGO detected another pulse of gravitational waves. However, this event differed in two ways from the earlier detections. The signal was weaker and it came from a source only 130 million

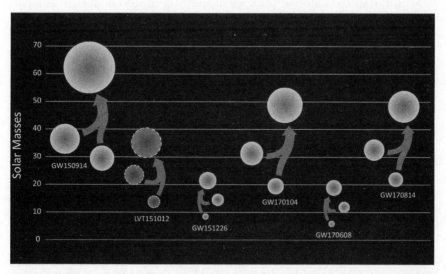

FIGURE 61. Detections from the first science run of the Advanced Laser Interferometer Gravitational-Wave Observatory (LIGO). The first detection occurred on September 14, 2015, just days after LIGO started taking science data after a long shutdown. There were two more detections in 2015, followed by three in 2017. The second detection (dashed circles) is considered only a candidate since its signal-to-noise ratio was below the threshold for a secure detection. *LIGO Scientific Collaboration*

light years away. That meant it was due to the merger of less massive objects, neutron stars instead of black holes.[43] LIGO worked alongside the European Virgo interferometer, and the signals from three different detectors enabled the scientists to pinpoint the gravitational waves with unprecedented accuracy. The neutron stars had collided in a galaxy called NGC 4993. The world's observatories swung into action.

The result was a bonanza of data and the birth of a new type of astronomy. Two NASA satellites detected a burst of gamma rays from the merging neutron stars, and over seventy telescopes across the globe caught the fading optical and infrared glow of the collision. Unlike black hole mergers, which produce no electromagnetic radiation, neutron stars combine in an explosion 1,000 times more powerful than a supernova. One result was the burst of radiation, and another was a flood of neutrons that powered a cloud of radioactive waste.[44] Within a day, the cloud expanded from the size of a city to the size of the Solar System. Neutrons impregnated atomic nuclei and built them into heavier elements. Theorists estimated that the event created 200 Earth masses of gold, worth about 10^{31} dollars if you could bring it home! The combination of gravitational waves and a rich harvest of electromagnetic information has been dubbed multi-messenger astronomy. It's expected that LIGO and Virgo will see about one neutron star merger every week, along with one black hole merger every two weeks.[45] The cosmos is awash with space-time ripples, and finally astronomers have eyes that can see them.

Accolades have followed swiftly. There's often a long time between a discovery and the subsequent award of a Nobel Prize. In fact, some eminent scientists have died waiting, and the prize cannot be awarded posthumously. But there was little doubt that detecting gravitational waves would be quickly recognized. Therefore, it was not a surprise when in October 2017, less than two years after LIGO felt its first shimmers in space-time, Rainer Weiss, Kip Thorne, and Barry Barish were named the winners of the Nobel Prize in Physics.

Collisions and Mergers of Massive Black Holes

Now that space-time ripples have been detected, here's what we can expect. There are a billion neutron stars and 300 million black holes in the Milky Way—plenty of candidates for mergers. However, the odds of them being in close binary systems are very low, so the black hole merger rate is about once every 500,000 years. That sounds like a lot of waiting around. But the sensitivity of LIGO gives it enormous reach across the universe. When advanced LIGO is back online in 2020 its sensitivity will be 3 times greater, which means that it can detect the same signal 3 times farther away. It will measure displacement to an incredible level of one part in 10^{22}. Since volume is proportional to the cube of the distance, the number of targets will go up by a factor of 30. The event rate could be as high as 1,000 per year, or a couple every day.[46]

The next regime will be to study gravitational waves emitted when the supermassive black hole at the center of a galaxy swallows a compact object like a neutron star or a black hole. To return to our sound analogy, the more massive the black holes, the longer the orbital time as they merge and the lower the frequency of the characteristic chirp. The supermassive object creates "sounds" in a frequency range of 10^{-4} Hz to 1 Hz and an orbital time of hours to seconds. Signals from a supermassive black hole would be below the range of human hearing, even below the lowest organ pipe; sounds more felt than heard.

Because their frequency range is so low, a detector must be located in the pristine environment of space in order to register gravitational waves from the most massive black holes. The proposed tool for the job is LISA, the Laser Interferometer Space Antenna. LISA will be a constellation of three satellites arranged in an equilateral triangle with separations of a million kilometers.[47] This configuration is 10 times the size of the Moon's orbit, and it will orbit the Sun at the same distance as the Earth, but trailing the Earth by 20 degrees. One satellite is the "master" with the laser and detector and the other two are "slaves" with reflectors attached to test masses made of an alloy of gold and platinum. LISA is designed to measure displacements smaller than the size of an atom over a distance of a million kilometers, or a precision of one part in 10^{21}. To detect tiny space-time ripples, the test masses must be immunized from any force other than gravity, as if they're not part of a spacecraft and are simply in the "free fall" of their Earth–Sun orbit. This engineering challenge requires exquisite control of the spacecraft. Each spacecraft has to float around its test mass, using capacitive sensors to determine its position relative to the mass, and precise thrusters to stay perfectly centered on that mass. In 2016, a test mission from the European Space Agency called LISA Pathfinder successfully demonstrated this technology. The success of LIGO led to a funding commitment in 2017, and the prospects for LISA are bright.[48]

In the standard model of cosmology, structures are built up hierarchically by the merger of smaller objects and accretion of surrounding matter. So dwarf galaxies combine to form large galaxies, and large galaxies keep growing by combining with the more abundant dwarf galaxies and by gas falling in from the intergalactic medium. The central black holes follow a similar buildup process, but details are difficult to predict because they depend on a complex accretion process and the particular conditions in the centers of galaxies.[49]

Mergers between supermassive black holes take place on even longer timescales and emit correspondingly lower-frequency gravitational waves. A rough calculation suggests that if a pair of million-solar-mass black holes merged they would emit gravitational waves with a frequency of 10^{-3} Hz and a timescale of an hour, while a merging pair of billion-solar-mass black holes would emit gravitational waves with a frequency of 10^{-9} Hz and a timescale of dozens of years. Catching a wave that takes years to pass through the detector requires extraordinary stability. Detailed computer simulations suggest that LISA will detect a few mergers per year, typically situations where both black holes are in the range 10^6 to 10^7 times the mass of the Sun.[50] This sampling range will give us perspective on the early phase of black hole and galaxy assembly.

However, the loudest events and the most spectacular mergers, of black holes a billion times the mass of the Sun, happen at frequencies so low they're beyond LISA's reach. To find these gravitational waves, an array a million kilometers across isn't large enough; an instrument on the scale of a galaxy is required. Enter the pulsar timing array. Pulsars are dead, collapsed stars made of pure neutrons. Hot spots on their surfaces sweep across radio telescopes as they spin, and the radio pulses keep perfect time. Pulsars that spin hundreds of times a second are the most accurate clocks in the universe.

Billions of light years away, two supermassive black holes do a leisurely dance that lasts millions of years. When they finally fall into each other's arms and merge, they bathe the universe in low-frequency gravitational waves that stretch and squeeze the fabric of space-time. Like us on Earth, the pulsars bob up and down as the waves pass by at the speed of light. The waves slightly alter the timing of the pulses. For example, a wave with a frequency of 10^{-8} Hz, or one cycle in four months, might cause the pulses to arrive 10 nanoseconds early in January and 10 nanoseconds late in March. It's an extremely delicate experiment, but current radio telescopes can measure pulses with the required precision. Arrays of pulsars are used to increase the sensitivity of the experiment and give some directional sensitivity.[51]

A pulsar timing array is the most grandiose experiment scientists have ever conceived. Rather than the 4 kilometers of LIGO or the million kilometers of LISA, pulsar detectors are arrayed over thousands of trillions of kilometers. The Milky Way galaxy is the detector. This is truly Big Science. There are four pulsar arrays actively looking for signals, and they are combining their data into an international array. As they add pulsars to their target lists and increase their sensitivity, there's

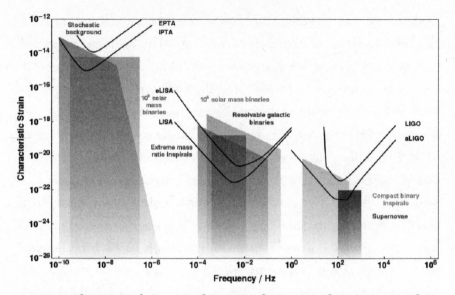

FIGURE 62. The various detectors and regimes of gravitational waves compared. To the right, at high frequency, are the sensitivity curves for interferometers like LIGO, which are sensitive to supernovae and compact neutron star and black hole binaries merging. In the middle are space-based interferometers like LISA, which detect lower-frequency events like massive black hole binaries merging. At the left are pulsar arrays, probing the lowest-frequency events, such as mergers of supermassive black holes and the stochastic background from the big bang. *C. Moore, R. Cole, and R. Berry, Classical and Quantum Gravity, vol. 32/Institute of Physics*

an 80% chance that one or more of these experiments will detect super-massive black hole mergers within the next decade (Figure 62).[52]

Gravity and the Big Bang

The wild frontier is the detection of primordial gravitational waves. Remember, ripples in space-time are created any time mass changes its motion of configuration. By far the most dramatic changes in mass took place in the early universe, when the matter that would eventually form hundreds of billions of galaxies was contained in a region smaller than an atom. Current cosmology includes an early phase called inflation: an exponential increase in the size of the universe when it was still micro-scopic, 10^{-35} seconds after the big bang. Inflation is invoked to explain the otherwise enigmatic flatness and smoothness of the universe.[53] Inflation implies that the "seeds" of galaxies were quantum fluctuations.

There's some indirect support for inflation, but the energy at that time was trillions of times higher than can be reached in the lab or in accelerators like the Large Hadron Collider so we cannot attempt to replicate it in a terrestrial experiment. Testing inflation is important because it will bring us closer to the "Holy Grail" of a theory of quantum gravity. Gravitational waves from inflation should still be reverber-ating through the universe. Their energy spreads across 29 powers of ten in frequency, encompassing all the detection methods we've dis-cussed.[54] However, the waves are too feeble to be measured with inter-ferometers or pulsar timing arrays, so astronomers have focused on their imprint on the radiation that bathed the universe as it cooled enough for stable atoms to form. That radiation has traveled unaltered through the universe since 400,000 years after the big bang, and we observe it as microwaves. The stretching and squeezing of space is predicted to leave a slight swirling pattern in the microwave radiation.[55]

The scientific community was electrified in 2014 when a team work-ing at the South Pole Telescope claimed to have detected gravitational waves caused by inflation—not directly, but by inferring them from their particular imprint on radiation.[56] The excitement was dashed a few months later when it turned out the team had been fooled by a contam-inating signal from dust in the Milky Way. It was a painful experience for the researchers, since they had checked the data carefully but were fooled by a subtle foreground, like having mist on your glasses and con-fusing it for a distant storm. The universe is a messy, complicated place that can't be controlled like lab equipment, so cosmologists are wise to be cautious. Yet when there's competition from other groups, the urge to publish quickly is hard to resist.

A number of teams are gearing up for new attempts at this important measurement. The best sites for these challenging microwave observations are near the South Pole and in the high and dry Atacama Desert in Chile. Five teams are involved. The stakes are high. If the gravitational wave signature isn't detected, a centerpiece of cosmology will be in doubt. But if the signature *is* detected, it will be direct evidence for quantum gravity.

A quantum origin for the universe may be a sign that we live in a multiverse, where we inhabit one out of a potentially infinite number of space-time bubbles. The universes in the multiverse are distinct space-times, probably unobservable from our space-time, which makes the idea difficult to test. They might all have different laws of physics and even be unrecognizably different from our universe. Do these other universes have the same fundamental forces? Do they contain black holes? Do they contain life forms that can understand their universe? These are some of the imponderable questions at the frontier of cosmology.

8.

THE FATE OF BLACK HOLES

THE FUTURE OF black holes is a story of short-term growth and long-term evaporation. Our distant descendants might witness the center of our own galaxy flaring up as a quasar, and the merger of the supermassive black holes in our galaxy and Andromeda. Eventually, black holes will reach a maximum size and no new black holes will be created. Life could persist in the universe even in the future era of darkness. But it would be severely challenged by the final victory of forces of dissipation and decay.

In the present, however, black holes present the ultimate test for any theory of gravity. The quest to reconcile the quantum theory with general relativity has led to gravity in multidimensional space-time. The three familiar dimensions of space only give hints of additional, hidden dimensions. Black holes must be incorporated into this new framework.

The New Age of Gravity

Why is gravity so weak? This doesn't seem like a sensible question, especially on a day when you find it hard to get out of bed—until, that is, you recall that a small bar magnet can hold up a paper clip against the downward-pulling gravity of the entire Earth. Gravity is vastly weaker than the other three fundamental forces; trying to explain this simple fact takes us down a rabbit hole of hidden dimensions and multiple universes.

As we've seen, physicists already have clues that the four fundamental forces may be manifested as a single super-force at sufficiently high temperatures or energies. The unification of two of the four forces was seen at accelerators in the 1970s, resulting in the award of a slew of Nobel Prizes. Pursuing this path led to the idea of supersymmetry. In the everyday world, particles with half integer spins, such as electrons and quarks (as a class, called fermions), don't interact with particles with integer spins that carry forces, such as photons and gluons (as a class, called bosons).[1] For subatomic particles, spin is an esoteric, mathematical property not directly analogous to the spinning of a top or a planet. Fermions and bosons are as aloof as oil and water. Supersymmetry unifies these categories by predicting a set of "shadow" particles for every fermion and boson, and it predicts that all the forces except gravity merge into one force at the phenomenal temperature of 10^{29} Kelvin. Theorists reached for supersymmetry to pursue their dream of unity underlying the plethora of different subatomic particles. But supersymmetry has been questioned, because no hint of these shadow particles has been seen with the Large Hadron Collider.

A second attack on unification emerged in the 1980s with string theory. String theory finesses the problems of the standard model of particle physics by hypothesizing that particles aren't fundamental but are oscillation modes of tiny one-dimensional entities called strings. Excitement about string theory spread like wildfire through the theoretical physics community. The theory was based on very elegant mathematics and it naturally united gravity with the other three forces. However, after more than a decade of intense research, many physicists grew discouraged with string theory. The mathematics was difficult and often intractable, and it required space-time to have nine dimensions, which seems like five too many! In string theory, the "hidden" dimensions are only realized at the incredibly high temperature of 10^{32} Kelvin or at the incredibly small scale of 10^{-35} meters. It looked as if the theory was untestable.[2]

Enter Lisa Randall. Growing up, she was drawn to math because it provided definitive answers. She was the first female captain of her school's math team and a classmate of noted string theorist Brian Greene at Stuyvesant High School in New York. As an eighteen-year-old, she won the Westinghouse Talent Search contest with a project about Gaussian integers. After getting her PhD at Harvard, she moved across the river to MIT as an assistant professor and became a rising star of theoretical physics.

Lisa Randall has a muse for music as well as math. There's not much opera inspired by theoretical physics. Even an opera buff might scratch his or her head, perhaps coming up with Philip Glass's *Einstein on the Beach*. Lisa Randall has added to this small oeuvre with *Hypermusic Prologue: a Projective Opera in Seven Planes*. Spanish composer Hector Parra wrote the score and Randall wrote the libretto.

To see why Lisa Randall was inspired to think creatively about gravity, let's return to the knotty problem of singularities. According to general relativity, every black hole contains a singularity, a place where the curvature of space-time is infinite.[3] Inside a black hole Einstein's equations fall flat on their face and predict something physically nonsensical. Stephen Hawking showed that singularities are inevitable in a black hole and he framed this problem dramatically: general relativity contains the seeds of its own destruction.

One possible path through this impasse involves string theory. String theory is motivated by a number of issues in fundamental physics. One is to unify the forces of nature in one framework. The "smooth" theory of curved space-time isn't consistent with the "grainy" theory of subatomic particles. This is the quest for quantum gravity that frustrated Einstein for decades. Also, the generally successful standard model of particle physics is flawed. Electrons have zero size in the model, so therefore must have infinite mass density and infinite charge density—another example of singularities seeming to violate physics. There's currently no explanation for why there are so many fundamental particles with different masses, or why matter is dominant over antimatter, or why dark matter and dark energy are the two major constituents of the universe.[4]

Randall knew that research on string theory in the 1990s had explored the richness of branes. "Brane" is an abbreviation for a membrane—a lower-dimensional object in a higher-dimensional space. Think of a sheet of paper, which is a two-dimensional object within three-dimensional space. Ants crawling on the sheet of paper are confined to move in two dimensions; they're unaware of the third dimension. There might even be another sheet of paper with ants crawling on it, and those ants would be unaware of the parallel "universe" not far away from them in a third dimension. Analogously, our own universe might be a brane, an island of three dimensions floating in a sea of higher dimensions. Particles are confined to the brane, but Randall knew that gravity wouldn't be confined to the brane because general relativity says it has to exist in the full geometry of space. She realized this might explain why gravity is so weak.

Randall had resisted the idea of extra dimensions for years, but at MIT she collaborated with Raman Sundrum of Boston University to brainstorm about branes. The math they came up with described a pair of universes, four-dimensional branes, thinly separated by five-dimensional space. They found that the space between the branes was warped and the warping could magnify and demagnify objects or forces between the branes. Therefore, gravity can be as strong as the other forces on one brane, but if we happen to be on the other brane we experience gravity as extremely weak (Figure 63). Randall and Sundrum were then stunned by another realization: that fifth dimension could be infinite and we'd be unaware of it. Until then, physicists assumed the conventional wisdom of

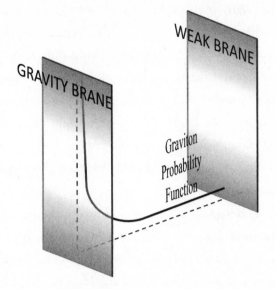

FIGURE 63. One possible explanation for the weakness of the gravity force involves branes or membranes: lower-dimensional objects embedded in a higher-dimensional space. Gravity might be strong on one brane but weak on another brane, both being three-dimensional spaces embedded in a five-dimensional space. It's not yet clear whether or not the higher dimensions or neighboring branes might be detectable in a lab or accelerator. *Chris Impey*

string theory, that the extra dimensions are coiled up so tightly that no experiment could probe them. In Randall and Sundrum's theory, they might be observable with accelerators.[5]

This work made them superstars. Sundrum got seven job offers. He mused about this good fortune, given how anxious he had been about their ideas, "Working that out was mind-blowing. We had reason to be dead scared. In each of these cases, there was a distinct fear of making complete fools of ourselves." Randall became the first tenured professor of theoretical physics in the long history of Harvard University. She ventured into writing popular books[6] and, less comfortably, has been regularly called on to speak for women in science. "I like to solve simple problems like extra dimensions in space," she noted wryly. "Everyone thinks women in science is a simpler issue, but it is so much more complicated."[7]

Branes are highly relevant for black holes. As we saw in chapter 1, Strominger and Vafa used string theory to reproduce the black hole entropy and radiation that Stephen Hawking derived using classical physics. By wrapping branes around tightly curled regions of space-time, theorists showed that they could calculate the mass and electric charge of the interior black hole. The fact that pure mathematics, developed for a completely different reason, can be used to calculate the properties of "real objects" like black holes was considered a triumph for string theory.

We could be living in a bubble of three dimensions floating in a sea of membranes of five, six, seven, or more dimensions.[8] All of these commingle in a construct called the multiverse. This is distinct from the multiverse described at the end of the last chapter, which is based on other space-times that may have emerged from quantum vacuum states existing alongside the big bang. The string theory multiverse is a set of shadowy multidimensional spaces that coexist with the universe we live in.

Higher dimensions haven't been detected yet in the laboratory or with accelerators, and many physicists think that branes, like strings, are clever mathematical constructs with little relation to reality. In some quarters, healthy skepticism has turned into a backlash. Yet Randall remains hopeful. The guru of gravity continues her work on the wild shores of higher mathematics. Let's leave the last word to a poet rather than a physicist, e. e. cummings: "listen: there's a hell of a good universe next door; let's go."[9]

Quasar on Our Doorstep

Black holes are evolutionary dead ends. For a massive star they represent the outcome where no energy can be generated and gravity is the victor. Supermassive black holes at the centers of galaxies are the deepest gravitational pits in the cosmos. They'll grow inexorably and they can't be starved forever. We have a ringside seat for the evolution of the nearest massive black hole, the one in our own galaxy. Is it possible to look back to a time when the Milky Way flared into life, to predict a time when it might burn brightly in the future?

The best probe of activity is X-ray emission, since it can reach us through the gas and dust of the galaxy's disk, whereas optical light is extinguished. In the twenty years that X-ray telescopes have been monitoring Sagittarius A*, it has mostly been very quiet. There are flares every few months that brighten it by a factor of 5 to 10 for less than an hour.[10]

But that's just twenty years of observation. It's possible to see changes in black hole fueling on timescales longer than a human life. Data combined from four different satellites detected the X-ray "echo" of a big flare from 300 years ago. At that time, Sagittarius A* brightened by a factor of a million, then the radiation reflected off a molecular cloud several hundred light years from the black hole before arriving at the Earth. The initial radiation mostly reached earth in the early eighteenth century, when no one had X-ray telescopes to observe it. The event itself occurred 27,000 years ago, when our early ancestors first reached northern Asia after leaving Africa for the first time.[11] An event this bright probably involved the black hole ingesting a star.

What about even longer timescales? Can we see what the currently slumbering black hole in the center of our galaxy was doing millions of years ago? Yes, and doing so solves a puzzle associated with the mass budget of the Milky Way. Our galaxy weighs a trillion times as much as the Sun. About 85% of that is dark matter—the invisible and mysterious substance that holds all galaxies together. That leaves about 150 billion solar masses of normal matter. Unfortunately, when astronomers added up all the stars, gas, and dust they could see it was half that amount. They found the missing matter with an X-ray telescope, in the form of hot dense fog permeating the galaxy. They saw a low-density "bubble" extending from the galactic center two-thirds of the way out to the Earth. They calculated the energy needed to evacuate such a large bubble and deduced that the Milky Way must have had a quasar phase in the past.[12] The shock wave is moving at a speed of 2 million miles per hour and it will reach us in about 3 million years, so there's no reason to panic. Tracing it back across 20,000 light years implies that the quasar phase started 6 million years ago, when early hominids walked the Earth. The timeline is corroborated by the presence of 6-million-year-old stars near the galactic center which probably formed from material that flowed toward the black hole during an even earlier feeding phase. The Milky Way black hole went on a feeding frenzy 6 million years ago, then belched out so much energy and gas that it ran out of food and went into hibernation.

What does the future hold for the galactic center? It's in a very quiet state now, but that won't last forever. We can expect the quasar on our doorstep to ignite every few hundred million years. There are signs that the galactic center is gathering itself for another active phase. X-ray observations have provided evidence for a swarm of 20,000 black holes and neutron stars within 3 light years of Sagittarius A$^\circ$.[13] This is the highest concentration of collapsed stellar remnants anywhere in the galaxy. They migrated to the center over a period of several billion years. If you have a bowl containing black marbles and wooden balls the same size and you shake it, the marbles will migrate to the bottom of the bowl because they're heavier. Similarly, gravitational interactions will cause black holes to be more centrally concentrated than the much larger number of normal stars.

Still, the odds are extremely low that we'll witness a return of quasar activity. For a galaxy like the Milky Way, the black hole is likely to brighten by a factor of a billion during only 1% of the remaining 5-billion-year lifetime of the Sun.[14] The last time the Milky Way was a quasar was when chimps and humans diverged on the evolutionary tree. The next time might be tens of millions of years in the future.

If we were still around as a species, what would we see? Nothing obvious. There's so much dust between us and Sagittarius A* that most visible light would be blocked. Radio jets, invisible to the human eye, would bisect the sky at right angles to the Milky Way. High-energy radiation would also surge, causing elevated mutation rates. Unless we took permanent shelter, our DNA would be steadily shredded. But if we could rise 100 light years out of the galaxy disk, from that perspective we'd have a majestic view onto the flaring black hole, its accretion disk shining as bright as a full Moon.

Merging with Andromeda

We're on a collision course with our nearest neighbor. Before the Sun dies, the Milky Way and Andromeda will approach, interact, and merge, with uncertain consequences for the Solar System and its inhabitants. The merger of the black holes at the center of each galaxy will be one of the most spectacular events imaginable.

We've known for a century that M31, the Andromeda galaxy, is approaching us at a speed of 120 kilometers per second, or 270,000 mph. Galaxies are generally receding from us due to the expansion of the universe, but the Milky Way and Andromeda are close enough that their mutual gravity overcomes the cosmic expansion. Hubble Space Telescope measurements of the sideways motion of Andromeda reveal that it's heading almost directly toward us.[15] Simulations show that in 2 billion years the galaxies will sweep past each other. As they move apart, a ghostly bridge of stars and gas will connect them. Currently, Andromeda is a faint, fuzzy patch of light, barely visible to the naked eye. Four billion years from now, Andromeda will loom large in the nighttime sky, for anyone on Earth who is here to see it (Figure 64). About 4.5 billion years from now the galaxies will approach each other again, execute a few tight loops, and merge. Over the following billion years they will settle into a smooth, large new galaxy: Milkomeda.

FIGURE 64. The Milky Way and Andromeda are on a collision course. This image shows the Earth's night sky nearly 4 billion years from now, after the two galaxies have swept past each other and are approaching for a final time. The Milky Way is distorted by the interaction. After the galaxies merge, the black holes at the center of each galaxy will also merge to produce a new, more massive black hole. *NASA/Space Telescope Science Institute*

This hypothetical new galaxy was named by Harvard University's Avi Loeb, who worked with Harvard postdoc T. J. Cox on computer simulations of the merger. They varied the assumptions and starting conditions, and each time it took two weeks on the equivalent of twenty state-of-the-art desktop computers.[16] A collision between two galaxies isn't like a car crash. Galaxies are mostly empty space, so very few stars actually hit one another. At the position of the Sun, if stars were the size of golf balls they'd be separated by 1,000 kilometers, and even in the center of the galaxy the separations would be 3 or 4 kilometers. Gravity will move stars around dramatically but their solar systems will remain intact, so future Earthlings will be treated to a new night sky as we're wrenched from our routine orbit of the Milky Way disk into a new situation.

What will happen to the Earth and the Solar System in this galactic train wreck? There's a 10% chance that the Sun will be flung into a tidal tail after the first close passage of the two galaxies. (A tidal tail occurs when the gravity of two extended objects disturbs and distorts them.) That would give us a bird's-eye view of the subsequent action. There's even a 3% chance that Andromeda will "steal" the Sun from the Milky Way. On the second and final approach, there's a 50% chance that the Sun will move toward the dense inner region of Milkomeda and a 50% chance that it will be cast out and our descendants will get to watch from afar as gravity blends the results of the collision into a smooth galaxy.

All this is an entertaining sideshow. The main event is the encounter between the 4-million-solar-mass black hole in the Milky Way and the 50-times-larger black hole in Andromeda.[17] The black holes will converge near the center of Milkomeda, moving inward by transferring energy to stars they encounter, some of which will be ejected from Milkomeda entirely. This will take about 10 million years. When they approach within a light year of each other, they will enter a death spiral and release a paroxysm of gravitational waves before merging.[18]

The merger between the Milky Way and Andromeda will not be unusual. Events like this are happening continuously in the universe. While the merger rate has declined as the universe expands, it's still significant. But not all mergers follow the pattern of the black holes measured by LIGO. When binary stellar black holes merge and their spins happen to be counteraligned, the gravitational waves can carry away enough momentum that the merged pair experiences a recoil "kick." The force of the kick can be enough to eject the merger remnant. Galaxies like the Milky Way occasionally spit black holes into intergalactic space. This can also happen to supermassive black holes after two galaxies merge. It's wonderful to imagine huge and naked black holes sailing through the space between galaxies at millions of miles per hour.

Half a dozen binary supermassive black holes have been identified. In the previous chapter we saw that the LISA space interferometer is designed to detect the merger of such binaries. The theoretical tools required to model these mergers have only been developed recently.[19]

We won't have to wait billions of years for a signal, as we would with Milkomeda. The quasar PG 1302-102 is 3.5 billion light years away. It has a binary black hole in a five-year orbit, indicating that the black holes are only a light month apart. That means the death spiral is imminent (although because of the time it takes the information to reach us, it actually happened 3.5 billion years ago). An even more exciting prospect is a pair of black holes 10 billion light years away, each of which is several billion times the mass of the Sun.[20] The orbital time of a year and a half means they're separated by 6 times the Schwarzschild radius, so the system is close enough to merging that it should be pouring out gravitational waves. These black holes actually merged billions of years ago, but we may only have to wait a few millennia to hear their space-time song.

The Biggest Black Holes in the Universe

Supermassive black holes bring to mind Gargantua, the dark centerpiece of the movie *Interstellar*. Gargantua is the destination of space travelers who hope to use a wormhole to vault across space-time. It's 100 million times the mass of the Sun, its event horizon is the size of the Earth's orbit, and it's spinning at 99% of the speed of light. As we've seen, Gargantua is the most realistic depiction of a black hole in popular media thanks to the input of Kip Thorne, who made sure the film did justice to both science and art.[21]

Gargantua is 25 times the mass of the black hole at the center of the Milky Way, but it's puny compared to the most massive black holes. The Sloan Digital Sky Survey located ten black holes over 10 billion solar masses in the distant universe.[22] They must have accreted material very rapidly to grow by a factor of a million from their seed mass in only 1.5 billion years. These behemoths dwarf the size of the Solar System (Figure 65). The record holder is a strong radio-emitting quasar with a black hole mass of 40 billion solar masses.[23]

FIGURE 65. The black hole at the center of the nearby galaxy NGC 1277 is potentially the most massive ever found, about 17 billion solar masses, though another study measured only 5 billion solar masses. This diagram shows the size of the event horizon in comparison to the Solar System. This galaxy has a black hole 10 times larger relative to its stellar mass than most galaxies. *D. Benningfield/K. Gebhardt/StarDate*

Astronomers are cavalier about tossing around large numbers, so let's pause to digest the implications of extreme black holes. A black hole 40 billion times more massive than the Sun has a Schwarzschild radius of 4 light days, so the event horizon is 20 times the size of the Solar System out to Pluto and the other dwarf planets. The black hole spins at a substantial fraction of the speed of light. Whereas the outer planets in our Solar System take 250 years to complete an orbit, this much larger object spins once every three months. Although the mass of a small galaxy has been squeezed into a Solar System volume, the mean density is 100 times less than the air you're breathing. The black hole emits no light but the surrounding accretion disk glows brightly. A black hole of this mass in an active quasar phase would emit 100 trillion times the luminosity of the Sun.

What lies ahead for the most massive black holes in the universe? Galaxies grow by accreting gas from the voids of space and by mergers. Both of these paths for growth are diminishing. As the universe gets larger, the gas supply thins out and galaxies get more widely separated so they merge less often. There's a correlation between stellar mass for a galaxy and its central black hole mass. It ranges from 10^4 to 10^5 solar mass black holes in globular clusters through 10^6 to 10^7 solar mass black holes in galaxies like the Milky Way up to 10^{10} solar mass black holes in elliptical galaxies with a trillion solar masses of stars. Regardless of the size of the stellar system, the central black hole is about 1% of the total mass in stars, and it's only 0.1% of the mass of the galaxy when dark matter is included.

I've spent years trying to understand the life and times of supermassive black holes. My student Jon Trump and I racked up dozens of nights on 6.5-meter telescopes in Arizona and Chile. Modern instrumentation means data that once took a lifetime to gather can be gathered in the time it takes for a graduate student to finish a thesis. With classical spectroscopy, the light from one active galaxy passes through a slit and gets spread into a spectrum. The instrument we were using in Chile placed little slits on hundreds of targets over an area of sky the size of a full Moon. A single long exposure might yield 100 black hole masses. With this data we hoped to tell the story of the rise and fall of quasar activity in the universe. Quasars are so far away that it makes no difference which direction you point the telescope, but I'm partial to the southern sky. The Milky Way arcing overhead like a ragged silver curtain is spectacular, and there's the bonus of our neighbor galaxies, the Magellanic Clouds, suspended like cotton wool balls on black cloth. It was so dark outside I could read a book by starlight.

We gathered statistics that traced the overall arc of black hole evolution over cosmic time. To do this meant sampling all the black holes, not just the extreme ones. I'd got over my youthful obsession with blazars and now wanted to know what controlled the overall population of active galaxies. By analogy, if you want to know the population of cars, you'll be counting a lot more Fords and Toyotas than Ferraris and Aston Martins. One big mystery was the fact that black holes are only active a small percentage of the time. Another is the tight relationship between the mass of the central black hole of a galaxy and the mass of all the old stars in the galaxy, which are distributed on much larger scales. It's as if the black hole "knows" what kind of galaxy it lives in.

In our data, the biggest black holes grew quickly in the first few billion years after the big bang, and then were starved of fuel. The more numerous smaller ones grew slowly until, in the past 5 billion years, they too became mostly quiet. The peak of the quasar era is long over, but black holes don't disappear, so presumably they're being "starved," with less fuel to power them as time goes by. That makes sense because the expanding universe is getting less dense and the rate of galaxy collisions is declining. But for any particular epoch in cosmic time and any particular galaxy mass, we cannot predict which black hole will be booming and which will be silent. It's equally hard to predict the future of these quasars.

We made the research into a game, laying out the quasars as cards on a table, like stamp collectors. Were some of them bright because they had a companion galaxy to feed on? In a few cases, but not always. Were some dim because they lived in a gas-poor galaxy? Not necessarily. We couldn't identify any trigger for nuclear activity. The overall painting made sense, but the individual dots of paint could be any color of the rainbow.

Nature is creative: it makes black holes that span a factor of a billion in mass (Figure 66). In our work, we've never found a black hole more than 10 billion times the Sun's mass. That's a little disappointing; I always wanted to claim that on my CV. Theorists predict a limit on black hole mass 10 times larger, about 10^{11} solar masses.[24] At that level, accre-

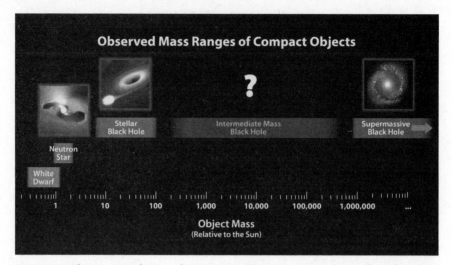

FIGURE 66. The masses of super-dense cosmic objects, ranging from white dwarfs to supermassive black holes in the nuclei of galaxies. The three types of object at low mass form when stars die, with more massive stars leaving more massive remnants. Only a few intermediate-mass black holes are known; they are found at the center of globular clusters or dwarf galaxies. The most massive black holes are found at the centers of the most massive galaxies in the universe. *NASA/JPL/California Institute of Technology*

tion physics becomes important, regardless of the host galaxy's mass. It seems to be nature's limit on black holes. To grow even bigger, the black hole would have to accrete 1,000 solar masses per year, and that much gas would collapse into new stars on scales of hundreds of light years, well before reaching the black hole. Also, the black hole starts to self-regulate. Radiation flooding out pushes away the incoming gas and stops further feeding. The bloated beast is grasping for food but there's none within reach.

The Era of Stellar Corpses

Even though the massive black holes at the centers of galaxies are approaching a natural limit, the death of massive stars continues to lead to more low-mass black holes. Stellar evolution is a battle between the forces of light and darkness—energy from nuclear fusion keeps the star puffed up while gravity tries to make it shrink. As we've seen, these forces will be balanced in the Sun for another 5 billion years, then gravity will win and crush the core into a white dwarf. Massive stars evolve more quickly, and when gravity wins they'll leave behind neutron stars or black holes.

The universe is heading for darkness. The first star formed about 100 million years after the big bang, when the universe was 30 times smaller and hotter than it is now. Galaxy assembly and star formation peaked about 3 billion years after the big bang and they've been declining ever since. Stars are currently being formed at a thirtieth of their peak rate and the decline will continue as less and less gas is available to form new stars. Even if we waited forever, only 5% more stars would form than have been formed up to this point.[25] These are averages; at any epoch more massive and gas-rich galaxies have higher star formation rates than less massive and gas-poor galaxies. The diminishing supply of gas will be eked out for a long time by stars that eject some of their mass late in life or die as supernovae.

Along with the declining rate of new stars forming, an increasing fraction of the stellar mass in all galaxies will be in the form of collapsed remnants. Once star formation finally ceases and the last black hole forms, roughly 100 trillion years from now, gravity will have its final victory.[26] By coincidence, this is the life expectancy of the lowest-mass red dwarfs, which are cool stars just above the mass sufficient to maintain fusion: 0.08 solar masses. The timescale is enormous. We're still in the very first phase of the universe as illuminated by stars, the equivalent of a baby only a week old.

In this distant future, as the stellar era ends, the 400 billion stars of Milkomeda will be equally divided between white dwarfs and brown dwarfs, with a small residue of neutron stars and black holes. Stars above 0.08 solar masses and below about 8 solar masses will collapse to roughly the size of the Earth and radiate their remaining energy into space as white dwarfs. Failed stars ranging from 0.08 solar mass down to 0.01 solar mass (from 10 to 80 Jupiter masses) will collapse into brown dwarfs, perhaps feebly fusing hydrogen into lithium.[27] Neutron stars will be 0.3% of the total stellar remnants in Milkomeda and black holes will be a paltry 0.03%.

As the eons pass, the white dwarfs and brown dwarfs will cool such that their radiation shifts to invisible infrared wavelengths. For a while, black holes in binary systems will be bright due to gas siphoned from their companions. But eventually the companions will also become stellar corpses and that source of gas will be exhausted. Galaxies will then slowly fade to black.

A Future of Evaporation and Decay

The far future we've just described applies not only to Milkomeda but to each of the hundreds of billions of galaxies in the observable universe. Their stars are subject to the same laws of astrophysics as stars in our system. But our descendants will never get to see all the other galaxies go dark. The reason is dark energy.

Dark energy is the biggest enigma in cosmology. Astronomers discovered in 1995 that the universal expansion is accelerating due to something that works in opposition to the force of gravity, which should cause deceleration. The cosmic "pie" is 25% dark matter, 70% dark energy, and 5% normal matter. Black holes large and small make up 0.005% of the universe, so they're a very minor component.[28] Dark energy means that galaxies we can see now will steadily be removed from view because they'll be receding faster than the speed of light. In 100 billion years, or 10 times the age of the universe to this point, all galaxies beyond Milkomeda will have exited our event horizon.[29] We'll be reduced to staring at our navel, metaphorically speaking. The end of the stellar era and subsequent events will only be measurable in the galaxy we inhabit now.

After Milkomeda goes dark, the future is evaporation and decay. Over time, stars within a galaxy exchange energy, with lighter stars tending to gain energy and heavier stars tending to lose energy. Recall the analogy of a bowl containing black marbles and wooden balls the same size where, if it's shaken, marbles migrate to the bottom of the bowl. Some stars will gain enough energy to leave Milkomeda, leaving the galaxy smaller and denser. This increases the interaction rate between stars so the process accelerates. At the same time, the decay of stellar orbits caused by the emission of gravitational radiation will move stars inward. After about 10^{19} years, 90% of the stellar remnants will have been ejected. Milkomeda will evaporate, with the remaining 10% of the remnants falling into the supermassive black hole. After the merger of the Milky Way with Andromeda the central black hole will be about 200 million times the mass of the Sun. It will eventually grow to about 10 billion solar masses.[30] If the age of the universe to this point were the first week of your life, you would need to live another 10 million years to watch this play out.

Thereafter, the distant future gets hazy and speculative. Physicists venture beyond the standard model of particle physics to explain why the universe contains far more matter than antimatter and to try and unify the electromagnetic force with the weak and strong nuclear forces. These schemes are called Grand Unified Theories (GUTs) and most of them predict proton decay. If protons decay, normal matter is not stable. Proton decay has never been observed and the current limit is 10^{34} years, which rules out some but not all Grand Unified Theories.[31] If protons decay, all stellar remnants except black holes will disintegrate into electrons, neutrinos, and photons.[32]

The final dissolution of the universe takes a staggering amount of time. Assuming that normal matter decays, only stellar and supermassive black holes will remain. Stephen Hawking predicted that black holes emit a feeble amount of low-energy radiation that causes them to slowly evaporate. It's important to recognize that this is speculation, as Hawking radiation has never been observed and no technology to detect it yet exists. The timescale for massive star remnants to evaporate is 10^{76} years. The supermassive black hole at the center of Milkomeda will evaporate in 10^{100} years. Everyday analogies fail miserably to convey this level of near-eternity. Even this is just a way station on the path to terminal heat death for the universe (Figure 67).

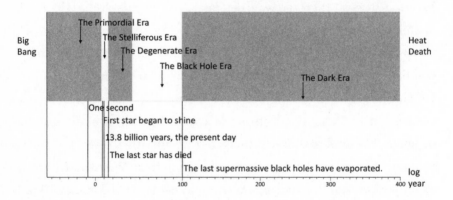

FIGURE 67. Far future timeline of the universe. On this logarithmic chart, the entire history of the universe up to this point is at the far left. Even the disappearance by evaporation of the most massive black holes is not the last physical process. After yet another vast interval, normal matter decays and the universe is left as a high-entropy soup of particles and low-energy photons. *Chris Impey*

"Things fall apart; the center cannot hold; mere anarchy is loosed upon the world." So wrote William Butler Yeats in 1919.[33] He was commenting on World War I but he might have been anticipating the end of the universe. The scientific context for this outcome is the second law of thermodynamics: the universal tendency to increasing entropy and disorder. Arthur Eddington confirmed general relativity but didn't believe in its prediction of black holes. However, on the inevitability of heat death, he was unequivocal. He wrote, "The law that *entropy* always increases, holds, I think, the supreme position among the *laws of Nature.* If someone points out to you that your pet theory of the universe is in disagreement with *Maxwell's equations*—then so much the worse for Maxwell's equations. If it is found to be contradicted by observation—well, these experimentalists do bungle things sometimes. But if your theory is found to be against the *second law of thermodynamics* I can give you no hope; there is nothing for it but to collapse in deepest humiliation."[34]

Black holes are enigmatic, so it's appropriate that they will be the last objects remaining at the end of the universe.

Living with Black Holes

Our narrative has become dark and gloomy; don't forget that the universe is built for life. Although astronomers haven't found any examples of biology beyond the Earth, they're optimistic that they will. In the Solar System, there are potentially habitable locations on Mars, Europa, and Titan, and on a dozen giant planet moons where water exists under a crust of rock and ice.[35] In 1995, the first exoplanet, or planet orbiting another star, was discovered after decades of fruitless searching. Since then the floodgates have opened and the current census of confirmed exoplanets stands at over 3,700.[36] The early exoplanets were discovered using the Doppler method, which revealed them because of the way they tugged on their parent star; more recently most discoveries have used the transit method, where the exoplanet eclipses and momentarily dims its parent star.[37]

The Milky Way contains a staggering 10 billion terrestrial planets with surface conditions suitable for liquid water.[38] Most of the 100 billion stars in the Milky Way have terrestrial planets. If life requires only carbon material, liquid water, and a local source of energy, there may be several hundred billion habitable locations on moons and planets where the surface is less hospitable. The real estate of time is as important as the real estate of space. There was enough carbon in the universe for an Earth "clone" to form within a billion years of the big bang, so some terrestrial planets have an evolutionary head start of up to 8 billion years on the Earth. Our ignorance is simply too great to imagine all the forms of biology that might evolve on these myriad worlds.

Given that we don't even know of life on one other world, it might seem presumptuous to ask about the prospects for life in the far future, but let's do it anyway.

Life doesn't need a star; it simply requires an energy source. According to the second law of thermodynamics, biology needs a temperature difference to provide a usable source of energy. Life on Earth exploits the temperature difference between the Sun and the cold vacuum of space. The Earth absorbs photons from the Sun at 6,000 Kelvin and emits 20 times more photons at 300 Kelvin into the sky. Biological organisms run complex processes that lower the entropy or disorder locally, but those organisms emit heat or waste energy that's eventually radiated into space. The energy argument applies even if life becomes computational (in the sense of AI) rather than biological, because any manipulation of information requires some form of energy.

As the stars in the universe exhaust their nuclear fuel, a hypothetical civilization of the far future could still exploit the temperature difference between cooling embers—white dwarfs and brown dwarfs—and deep space. The physicist Freeman Dyson considered the future of life and concluded that biology could be sustained in an era of diminishing returns by hibernating for increasingly long spans.[39] That will work for 10 billion years or so, but what about when all the stars fade to black?

Salvation may come from black holes. Energy could potentially be extracted from a black hole's spin. Just beyond the event horizon is a region called the ergosphere. The word comes from the Greek for "work" and it was coined—no surprise here—by John Wheeler. The ergosphere is dragged around by a spinning black hole like water is dragged around by a whirlpool, and it's thinner at the black hole's poles—think of a spinning balloon filled with water where rotation causes it to bulge at the equator. Roger Penrose proposed in 1969 that it was possible to extract energy from the ergosphere.[40] With the right trajectory, an object can enter the ergosphere and leave with more energy than it had on the way in. The black hole will spin very slightly slower as a result. A civilization could do careful calculations, toss things into a black hole, and harvest the extra energy they have when they're ejected again.

Another clever idea is to flip the temperatures around and have a cold star and a hot sky. Black holes in the present universe are often bright because matter falls toward them and forms a hot accretion disk. However, in the far future the gas will have been consumed and black holes will be cold and dark, except for a feeble drizzle of Hawking radiation with a temperature of a fraction of a degree. Compared to that, the universe has a balmy temperature of 2.7 Kelvin due to radiation left over from the big bang, a temperature that will decline as the universe continues to expand. Theorists calculate that an Earth-like planet orbiting a black hole close enough that it appears the same size as the Sun in our sky could extract about a kilowatt from the temperature difference.[41] It's probably enough to sustain a miniature or very efficient civilization (Figure 68).

A similar strategy was used in the film *Interstellar*, in which a world called Miller's planet orbits close to the massive, spinning black hole Gargantua. Gravity slows down time so much that during one hour on the planet seven years pass off-world. In this scenario, the inhabitants of Miller's planet can generate 130 gigawatts of power, but the film is off the mark in imagining people could live there. Such a large amount of energy heats the planet to 900 degrees Celsius, enough to melt metal.

The problem with using black holes as a means to draw energy from big bang radiation is the speed of the cosmic expansion. The temperature of that radiation is 2.7 Kelvin now, but as dark energy causes the universe to grow exponentially those photons are stretched by the

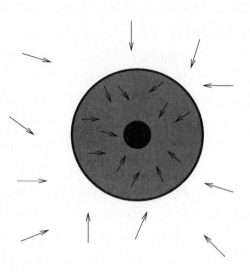

FIGURE 68. A method for a far future civilization to extract a small amount of energy from a black hole. In a conventional Dyson sphere, a shell built around a star captures energy from the star and radiates waste heat outside. In this version, Hawking radiation from the black hole is colder than microwave radiation from the big bang, so the shell absorbs microwave radiation from the outside and radiates waste heat to the black hole, leaving a little energy to be harvested. *T. Opatrny, L. Richterek, and P. Bakala, Am.J.Phys., vol. 85/American Institute of Physics*

expansion to have very long wavelength and low energy. Within 100 billion years, the big bang radiation will have a temperature that's a tiny fraction of a degree.

Civilizations will have to switch strategies. The Hawking radiation for a minimum-mass black hole 3 times the mass of the Sun has a temperature of 2×10^{-8} Kelvin and a luminosity of 10^{-29} watts. That's truly feeble, but apart from black hole spin it will be the only energy available until these black holes evaporate after 10^{76} years. To capture all the radiation, a civilization would have to surround the black hole with the kind of sphere Freeman Dyson imagined intelligent aliens might use.[42] Then attention will turn to the massive black hole at the center of Milkomeda. With a temperature of 6×10^{-18} Kelvin and a luminosity of 10^{-48} watts, it's a feeble fire at which to imagine warming one's hands. Living in the far future will require parsimony and patience. But until that last black hole evaporates after 10^{100} years, time is one thing in the universe that will never be in short supply.

I have glimpsed black holes in my research. They are massive and inscrutable, seen across the gulfs of space to distant galaxies. My span will be short compared to theirs. How long will they endure? Blink quickly. You could have done that a billion billion times since the big bang. The time until the most massive black holes dissipate is to the age of the universe as the age of the universe is to the blink of an eye. And so on three more times, to reach 10^{100} years.

This much time is unfathomable. The word "clock" is quaint. It comes from the Middle English word for bell, a reminder of a time when clocks had no hands and no numbers because so few people could read. Long after human time, after pendulum clocks, after the mechanical time of Timex and Rolex, after the last radioactive atoms have decayed, and after the last pulsar has spun down, there will be black hole time.

I imagine I'm immortal. If I could watch until the end of black hole time, and see what we or civilizations from other stars have done, what would I see?

First, there is a Barbaric Age, an extension of the one we live in, when civilizations rage against one another and the worst fate for a vanquished foe is to be flung into a black hole and suffer the torments of being torn asunder by gravity. Then perhaps, the Civilized Age, when creatures leave images frozen onto the event horizons of large black holes as timeless memorials. As an optimist, I imagine an Age of Knowledge, when some learn how to read the information stored holographically on the event horizon, and others venture into spinning black holes to take refuge on the time-like surface, a hall of temporal mirrors where you can travel back and forth to meet your past and future self, but never leave. Finally, an Age of Sentience, when life is distilled into pure computation and black holes are a form of information storage. How pleasing to think that these ciphers might sustain the heartbeat of the universe.

Gravity is the weakest force, but it's the most grandiloquent and the most persistent. The other forces have long since quit. Subatomic particles have all decayed and electromagnetic radiation has been diluted and stretched into oblivion. All the crashing chords of gravitational radiation as black holes merge are in the past. The only music of the spheres is the low thrum of black holes spinning. Slowly, inexorably, they evaporate. This is the end. The universe has dissipated to an almost perfect smoothness, with the vacuum lightly ruffled by quantum fluctuations.

NOTES

FOREWORD

1   The phrase also alludes to a collection of short stories by the British writer Martin Amis. The stories dwell on the threat of nuclear war, and the allusion is to $E = mc^2$, the equation in which Einstein pointed to the enormous power of the atomic nucleus. See Martin Amis, *Einstein's Monsters* (London: Jonathan Cape, 1987).

1: THE HEART OF DARKNESS

1   R. MacCormmach, *Weighing the World: The Reverend John Michell of Thornhill* (Berlin: Springer, 2012).

2   J. Michell, *Philosophical Transactions of the Royal Society of London* 74 (1784): 35–57.

3   S. Schaffer, "John Michell and Black Holes," *Journal for the History of Astronomy* 10 (1979): 42–43.

4   The Michelson–Morley experiment was an attempt to detect the aether, a diffuse medium pervading space that was hypothesized to carry the force of gravity and mediate electromagnetic waves. This famous "failed" physics experiment found that light arrives at the same speed, regardless of the Earth's 30 kilometer-per-second motion around the Sun. The null result of this experiment was pivotal in framing the special theory of relativity. Recent data rules out the presence of a light-carrying medium at a level of one part in 10^{17}.

5   C. Montgomery, W. Orchiston, and I. Whittington, "Michell, Laplace, and the Origin of the Black Hole Concept," *Journal of Astronomical History and Heritage* 12 (2009): 90–96.

6   When I was a student studying physics in London, I visited Cambridge to try and get a measure of Isaac Newton. I wanted to understand the man behind

the equations. With help from a colleague I got access to Newton's rooms at Trinity College. His study had narrow arched windows and was lined with dark wood, so it was gloomy even at noon. I'd read that he solved problems by "thinking upon them without ceasing," and my guide told me a story of one of the rare times Newton entertained guests. Going to the back room to get a bottle of port, he saw an unfinished calculation on his desk and sat down to complete it. His forgotten guests quietly let themselves out. Out in the quadrangle, I walked on the gravel paths where, 300 years earlier, Newton drew scientific diagrams with a stick. The Fellows of the College learned to sidestep them in case they interfered with a work of genius. That afternoon I drove to Newton's childhood home at Woolsthorpe Manor. As a youth, he was often sent into the nearby village on errands or to take the family horse to be shod. His mother would find him hours later standing on a bridge, staring at the water, lost in thought, the errands forgotten and the horse having slipped its traces. I was pleased to see an apple orchard behind the house.

7   From the preface to Richard S. Westfall, *Never at Rest: A Biography of Isaac Newton* (Cambridge, UK: Cambridge University Press, 1983).

8   J. Stachel et al., *Einstein's Miraculous Year: Five Papers That Changed the Face of Physics* (Princeton: Princeton University Press, 1998).

9   Thought experiments are powerful tools for advancing science. Dating back to ancient Greek philosophy, they are a way of posing a hypothetical question to Nature. Galileo provides an early example in physics, when he talks about dropping different objects from a tower to see their rates of descent (contrary to popular belief, he never actually did this experiment). Einstein used thought experiments to frame the issues of relativity, and physicists in the early twentieth century frequently used them to try and understand the implications of the quantum theory of matter.

10   The theory is mathematical and intimidating, but there are a number of popular or semi-technical introductions. Among the best are R. Geroch, *General Relativity from A to B* (Chicago: University of Chicago Press, 1978); D. Mermin, *It's About Time: Understanding Einstein's Relativity* (Princeton: University of Princeton Press, 2005); and of course the classic by Albert Einstein, *Relativity: The Special and General Theory* (New York: Crown, 1960). For a biography of Einstein, see A. Pais, *Subtle is the Lord: The Science and Life of Albert Einstein* (Oxford: Oxford University Press, 1982).

11   *The Sonnets of Robert Frost*, edited by J. M. Heley (Manhattan, KS: Kansas State University, 1970).

12   D. E. Lebach et al., "Measurement of the Solar Gravitational Deflection of Radio Waves Using Very-Long-Baseline Interferometry," *Physical Review Letters* 75 (1995): 1439–42.

13   C. W. Chou, D. B. Hume, T. Rosenband, and D. J. Wineland, "Optical Clocks and Relativity," *Science* 329 (2010): 1630–33.

14   N. Ashby, "Relativity and the Global Positioning System," *Physics Today*, May 2002, 41–47.

15   Quoted in S. Chandrasekhar, "The General Theory of Relativity: Why Is It

Probably the Most Beautiful of All Existing Theories," *Journal of Astrophysics and Astronomy* 5 (1984): 3–11.

16   I struggled with general relativity as a graduate student, and the experience convinced me that my future lay in observation rather than theory. Many years later I spent some time in Einstein's shadow while on sabbatical in Princeton. He spent nearly twenty years there, from 1936 until his death, working not at Princeton University but at the nearby Institute for Advanced Study. I once poked my head into his old office, apologizing to its current occupant, noted Canadian mathematician Robert Langlands. Walking from my rented house to the Institute, I passed Einstein's white clapboard house on Mercer Street. His house was later occupied by the physicist Frank Wikczek and then the economist Eric Maskin, both also Nobel Prize winners. I wondered if living in a house with that lineage would make you smarter. After Einstein died, his remains vanished. The autopsy surgeon removed Einstein's brain and stored parts of it in a jar in his office in Weston, Missouri. An ophthalmologist removed his eyes and stored them in a bank vault. In Princeton, I'd heard the rumor that his ashes were strewn into the Delaware River south of town. I ran along the riverbank and mused on the convoluted paths through space and time that had taken his atoms from the big bang, cycled them through the cores of stars, brought them together briefly for the singular insights of relativity, and then dispersed them to the sea.

17   *The Collected Papers of Albert Einstein*, volume 8A, *The Berlin Years: Correspondence*, edited by R. Schulmann, A. J. Kox, M. Janssen, and J. Illy (Princeton: Princeton University Press, 1999).

18   A. Pais, *J. Robert Oppenheimer: A Life* (Oxford: Oxford University Press, 2006).

19   J. R. Oppenheimer and H. Snyder, "On Continued Gravitational Contraction," *Physical Review* 56 (1939): 455–59.

20   R. Rhodes, *The Making of the Atomic Bomb* (New York: Simon & Schuster, 1986).

21   J. A. Hijaya, "The Gita of Robert Oppenheimer," *Proceedings of the American Philosophical Society* 144, no. 2 (2000), https://amphilsoc.org/publications/proceedings/v/144/n/2.

22   C. W. Misner, K. S. Thorne, and J. A. Wheeler, *Gravitation* (New York: W. H. Freeman, 1973).

23   A. Finkbeiner, "Johnny and Oppie," 2013, http://www.lastwordonnothing.com/2013/08/21/6348/.

24   Several excellent books have discussed Oppenheimer's complex feelings about his work on the bomb, and his fall from grace. See K. Bird and M. J. Sherwin, *American Prometheus: The Triumph and Tragedy of J. Robert Oppenheimer* (New York: Alfred A. Knopf, 2005), and M. Wolverton, *A Life in Twilight: The Final Years of J. Robert Oppenheimer* (New York: St. Martin's Press, 2008). An inside account of the atomic bomb project is H. Bethe, *The Road from Los Alamos* (New York: Springer, 1968). Many physicists were particularly bitter toward Edward Teller, who was more hawkish

than Wheeler and who pointedly failed to support Oppenheimer when he was stripped of his security clearance.

25   Quoted secondhand in Wheeler's autobiography: J. A. Wheeler, *Geons, Black Holes, and Quantum Foam: A Life in Physics* (New York: Norton, 1998).

26   In fact, the story is more complicated. Research by Marcia Bartusiak showed that the term "black hole" was first used at a scientific meeting in late 1963 and first appeared in print in early 1964. It is, however, indisputable that the term spread due to Wheeler's reputation. See https://www.sciencenews.org/blog/context/50-years-later-it's-hard-say-who-named-black-holes.

27   S. Hawking, *A Brief History of Time* (New York: Bantam, 1988). Hawking noted that the publisher told him that for every equation in the book, readership would be cut in half. So he pruned math from the initial manuscript down to the single equation $E = mc^2$. Nevertheless, the book is quite dense, so he followed up with a shorter, simplified version: S. Hawking, *The Illustrated Brief History of Time* (New York: Bantam, 1996). Carl Sagan's introduction to the first edition tells the story of an accidental encounter in London in 1974, as Hawking was being inducted into the Royal Society. As he watched the young man in a wheelchair slowly signing his name in a book that had Newton in its earliest pages, he realized Hawking was a legend even then.

28   Stephen Hawking was often reduced in the popular culture to an archetype—a brilliant intellect trapped in a wasting body—so it's difficult to get a sense of him as a person. Fleshing out the third dimension leads to some uncomfortable truths. His first wife, Jane Wilde, sacrificed her own academic career to care for Stephen and raise their three children, with minimal assistance. He later left her to live with one of his nurses (whom he married and then divorced). Wilde's memoir paints a picture of a man who could be an egotist and a misogynist, but her perspective has been subsumed by his written accounts and media treatments of the physicist that adhere to the heroic narrative. The edges to his personality don't mitigate his remarkable good spirits in the face of a lifelong debilitating illness. See Jane Hawking, *Music to Move the Stars: A Life with Stephen Hawking* (Philadelphia: Trans-Atlantic, 1999), and her second, softer version of the story, *Travelling to Infinity: My Life with Stephen* (London: Alma, 2007).

29   K. Ferguson, *Stephen Hawking: His Life and Work* (New York: St. Martin's Press, 2011). Older, but better on his contributions to physics, is M. White and J. Gribbin, *Stephen Hawking: A Life in Science* (Washington, DC: National Academies Press, 2002).

30   Euclidean geometry is the familiar formalism that applies to the linear space of Newtonian gravity. To come up with general relativity, Einstein reached into the toolbox of topology, a field of mathematics that describes space (with an arbitrary number of dimensions) that is deformed by stretching, crumpling, or bending. Part of his genius was to realize that the mathematics could be incorporated into a physical theory of gravity.

31   S. Hawking and R. Penrose, "The Singularities of Gravitational Collapse and Cosmology," *Proceedings of the Royal Society* A 324 (1970): 539–48.

32 Electric charge is a third possible property of a black hole. However, since black holes form from the collapse of matter that's electrically neutral, a charged black hole is considered artificial and unlikely. The electrical force is forty orders of magnitude stronger than gravity, so even a slight electric charge would stop a black hole from forming. Roy Kerr generalized the solution of a black hole to the spinning case, nearly fifty years after Schwarzschild's first solution, in R. P. Kerr, "Gravitational Field of a Spinning Mass as an Example of Algebraically Special Metrics," *Physical Review Letters* 11 (1963): 237–38. General relativity allows for such complex geometries of space-time that the equations can rarely be solved fully, and they can only be solved approximately by making strong assumptions about symmetry.

33 J. D. Bekenstein, "Black Holes and Entropy," *Physical Review D* 7 (1973): 2333–46.

34 S. Hawking and R. Penrose, *The Nature of Space and Time* (Princeton: Princeton University Press, 2010), 26. Hawking wrote many highly technical papers on black hole radiation and evaporation, but a slightly more accessible paper is S. Hawking, "Black Hole Explosions?" *Nature* 248 (1974): 31–32.

35 A. Einstein and N. Rosen, "The Particle Problem in the General Theory of Relativity," *Physical Review Letters* 48 (1935): 73–77.

36 S. Weinberg, *The First Three Minutes* (New York: Basic Books, 1988), 131.

37 M. Amis, *Night Train* (New York: Vintage, 1999), 114.

38 A. Z. Capri, *From Quanta to Quarks: More Anecdotal History of Physics* (Hackensack, NJ: World Scientific, 2007).

39 The colloquial meaning of entropy is disorder, but the original definition from physics is related to the number of equivalent microscopic configurations of a system. Since there are a huge number of ways to make a black hole, compared to the fairly limited number of ways to make a star, black hole entropy is very high. Mathematically, a black hole the mass of the Sun has an entropy 100 million times higher than that of the Sun.

40 D. Overbye, "About Those Fearsome Black Holes? Never Mind," *New York Times*, July 22, 2004, http://www.nytimes.com/learning/students/pop/2004 0723snapfriday.html.

41 This is a nod to Einstein, who called his alteration of a general relativity solution to match the astronomers' early 1900s' description of a static universe his "biggest blunder." Einstein added a term called the cosmological constant to counter gravity. Ironically, the universe is now known to be accelerating and that behavior is well described by the cosmological constant.

42 We're very familiar with tidal forces that operate in the Solar System. The near side of the Earth experiences stronger gravity from the Moon than the far side of the Earth, and when the oceans respond to this difference it creates the tides. The Sun also exerts a tidal force on the Earth, smaller because of its larger distance. When the tidal force on a solid object like a moon or an asteroid exceeds the strength of the rock, the object breaks up. This location is called the Roche limit. Tidal forces on Jupiter's small moon make it the most active volcanic world in the Solar System. Mathematically,

the tidal acceleration across an object of size d a distance R from an object of mass M is $2GMd/R^3$.

43   Scientific wagers have an intriguing history. One of the first known also involved gravity. In 1684, the English architect Christopher Wren offered a book worth two pounds (equivalent to $400 today) to anyone who could deduce Kepler's laws of planetary motions from an inverse square law for gravity. His wager was a deliberate effort to spur Isaac Newton into completing the calculation and publishing the result, which he later did in his masterwork on gravity, *Philosophiae Naturalis Principia Mathematica*. Newton missed the deadline for the bet.

44   A. Strominger and C. Vafa, "Microscopic Origin of the Bekenstein–Hawking Entropy," *Physical Letters B* 379 (1996): 99–104.

45   Reconciling quantum theory with general relativity consumed the last twenty years of Einstein's life. He was never successful. Some of the most obvious ideas about quantum gravity, such as that gravity is carried by a particle called the graviton, quickly run into technical problems. The role of time is also very different in quantum mechanics and in general relativity. String theory is considered a promising approach, but it generates a vast number of vacuum states that are difficult to sort through. Ironically, some of the recent progress made describing black holes with string theory involves switching off gravity! It will probably be a number of years before this research matures or generates predictions that can be tested.

46   A. Strominger and S. Hawking, "Soft Hair on Black Holes," *Physical Review Letters* 116 (2016): 231301–11. A more digestible interview with Andy Strominger about this work is in Seth Fletcher's blog *Dark Star Diaries*, http://blogs.scientificamerican.com/dark-star-diaries/stephen-hawking-s-new-black-hole-paper-translated-an-interview-with-co-author-andrew-strominger/.

2: BLACK HOLES FROM STAR DEATH

1   The pressure balance within a star getting its energy from nuclear fusion is called hydrostatic equilibrium. The process has negative feedback and acts like a thermostat. If for some reason the Sun felt pressure from outside and was squeezed, the temperature of the denser gas would go up, the nuclear reaction rate would go up, and more pressure would be created, expanding the Sun slightly. If for some reason the Sun expanded slightly, the interior temperature would go down, the nuclear reactions rate would also go down, and with less pressure created the Sun would shrink slightly. Stars like the Sun are long-term stable and nothing like bombs.

2   A star fusing hydrogen into helium is said to be on the main sequence. In the early twentieth century, astronomers Ejnar Hertzsprung and Henry Norris Russell showed that when the luminosity of a star is plotted against its color or surface temperature, it doesn't occupy all parts of that diagram. Most stars fall on a diagonal running from high luminosity and high temperature

to low luminosity and low temperature. Stars fusing other nuclear fuels or stars that have collapsed to their end states lie in other parts of the diagram.

3   The radiation law that governs stars is called the Stefan–Boltzmann law. It describes a blackbody, an object that is in equilibrium and has a constant temperature. The law says the total power radiated by a star is proportional to the product of the surface area and the temperature to the fourth power. So radiation emitted goes down quickly with decreasing size and even quicker with decreasing temperature.

4   E. Öpik, "The Densities of Visual Binary Stars," *Astrophysical Journal* 44 (1916): 292–302.

5   A. S. Eddington, *Stars and Atoms* (Oxford: Clarendon Press, 1927), 50.

6   Quoted in J. Waller, *Einstein's Luck* (Oxford: Oxford University Press, 2002).

7   The physical state of a white dwarf is called degenerate matter. Degeneracy pressure depends only on density, not on temperature. Degenerate matter is compressible, so the radius of a high-mass white dwarf is smaller and its density is higher than those of a low-mass white dwarf. The carbon-rich nature and quasi-crystalline atomic structure of white dwarfs led the rock group Pink Floyd to allude to them (and to their founding member Syd Barrett) in the song "Shine On, You Crazy Diamond," on their 1975 album *Wish You Were Here*.

8   S. Chandrasekhar, "The Maximum Mass of Ideal White Dwarfs" *Astrophysical Journal* 74 (1931): 81–82.

9   J. R. Oppenheimer and G. M. Volkoff, "On Massive Neutron Cores," *Physical Review* 55 (1939): 374–81.

10 P. Haensel, A. Y. Potekhin, and D. G. Yakovlev, *Neutron Stars* (Berlin: Springer, 2007).

11 Robert Forward accepted the challenge with *Dragon's Egg* (New York: Del Rey, 1980), now considered a classic of hard science fiction. He imagined tiny intelligent creatures that could live on the surface of a neutron star, with development and thought timescales a million times faster than humans.

12 See J. Emspak, "Are the Nobel Prizes Missing Female Scientists?" *Live-Science*, October 5, 2016, http://www.livescience.com/56390-nobel-prizes-missing-female-scientists.html. Women have fared only slightly better with other Nobel Prize subjects. Astronomy has gradually improved its gender balance, but men still outnumber women in the highest academic ranks and they win the lion's share of the major awards. I know Jocelyn Bell fairly well; we overlapped for a while at the Royal Observatory in Edinburgh and she and my mother went to the same Quaker meeting house for years. She vividly remembers the moment of discovery, seeing metronomic squiggles on a strip chart that had no obvious explanation. She played sleuth, tracking down and ruling out other explanations one by one. As for the Nobel Prize, there's no hint of bitterness when she talks about that early omission, and she's had an illustrious career in every other way. For her own story, see J. S. Bell Burnell, "Little Green Men, White Dwarfs, or Pulsars?" *Annals of the New York Academy of Science* 302 (1977): 685–89.

13 N. N. Taleb, *The Black Swan: The Impact of the Highly Improbable* (Lon-

don: Penguin, 2007). In this case, the black swan represents black holes, which had been predicted but were expected to be rare, and which some people thought could never be detected.

14   S. Bowyer, E. T. Byram, T. A. Chubb, and H. Friedman, "Cosmic X-Ray Sources," *Science* 147 (1964): 394–98.

15   The two papers that established Cygnus X-1 as the first viable black hole candidate were B. L. Webster and P. Murdin, "Cygnus X-1: A Spectroscopic Binary with a Massive Companion?" *Nature* 235 (1971): 37–38, and C. T. Bolton, "Identification of Cygnus X-1 with HDE 226868," *Nature* 235 (1971): 271–73. The paper that provided an accurate radio position for the X-ray source was L. L. E. Braes and G. K. Miley, "Detection of Radio Emission from Cygnus X-1," *Nature* 232 (1971): 246.

16   From Bruce Rolston, "The First Black Hole," news release, University of Toronto, November 10, 1997, https://web.archive.org/web/20080307181205/http://www.news.utoronto.ca/bin/bulletin/nov10_97/art4.htm.

17   The Canadian progressive rock band Rush heard about the first black hole soon after its discovery and wrote a song cycle called "Cygnus X-1," which featured on two of their albums, in 1977 and 1978. In this allegorical work, the explorer ventures into the black hole, crying out with a cry of "Sound and fury drown my heart. Every nerve is torn apart." In the second part of the cycle he's beyond the event horizon in a world called Olympus, where he reconciles the warring tribes of Apollo, who are ruled by logic, and Dionysus, who are ruled by emotion. The apotheosis of astronomy and rock was two years earlier. In 1975, Pink Floyd released their concept album *Wish You Were Here*, featuring the nine-part composition "Shine On You Crazy Diamond." The song is a double metaphor, on the one hand paying tribute to a man who shone brightly but flamed out while he was still young, and on the other hand alluding to white dwarfs as quasi-crystalline carbon. "There's a look in your eyes like black holes in the sky," Roger Waters sang.

18   It's similar to the situation with a seesaw or teeter-totter. When two people of equal weight sit at either end, they are balanced. With an adult and a child, the adult has to sit nearer the pivot to balance the child. This lever arm with a center of balance behaves like an orbit with a center of mass. When the masses are extremely unequal, like a planet orbiting a star, the star orbit is so subtle that it just wobbles. For example, Jupiter, the most massive planet in the Solar System, causes the Sun to wobble around its edge with a period equal to the 12 years of Jupiter's orbital period.

19   Orbits are generally elliptical rather than circular, but that complication doesn't affect the main argument. The velocity changes through the orbit, but the average velocity is the same as for a circular orbit of the same size.

20   The full solution to a binary orbit gives the equation $PK^3/2\pi G = M \sin^3 i/(1+q)^2$, where P is the period, K is half the full amplitude of the radial velocity variation, M is the black hole mass, and q is the ratio of the companion mass to the black hole mass.

21   D. Sobel, *The Glass Universe: How the Ladies of the Harvard Observatory Took the Measure of the Stars* (New York: Viking, 2016).

22   C. Brocksopp, A. E. Tarasov, V. M. Lyuty, and P. Roche, "An Improved Orbital Ephemeris for Cygnus X-1," *Astronomy and Astrophysics* 343 (1998): 861–64.

23   J. Ziolkowski, "Evolutionary Constraints on the Masses of the Components of the HDE 226868/Cygnus X-1 Binary System," *Monthly Notices of the Royal Astronomical Society* 358 (2005): 851–59.

24   J. A. Orosz et al., "The Mass of the Black Hole in Cygnus X-1," *Astrophysical Journal* 724 (2011): 84–95.

25   This brief discussion elides dozens of papers and thousands of hours of observations that elevated Cygnus X-1 to the status of a gold-plated black hole candidate. It took years for the observational errors to be reduced and other models to be ruled out. For example, as a way of avoiding the black hole inference, early models invoked a triple star system, with a blue supergiant and close binary consisting of a main sequence star and a neutron star. These models were eventually found to be highly unlikely. See H. L. Shipman, "The Implausible History of Triple Star Models for Cygnus X-1: Evidence for a Black Hole," *Astrophysical Letters* 16 (1975): 9–12.

26   J. Ziolkowski, "Black Hole Candidates," in *Vulcano Workshop 2002, Frontier Objects in Astrophysics and Particle Physics*, edited by F. Giovanelli and G. Mannocchi (Bologna: Italian Physical Society, 2003), 49–56, and J. E. McLintock and R. A. Remillard, "Black Hole Binaries," in *Compact Stellar X-Ray Sources*, edited by W. H. G. Lewin and M. van der Klis (Cambridge, UK: Cambridge University Press, 2006), 157–214.

27   There's another way to detect isolated black holes, based on the fact that they can pull in tenuous gas from the interstellar medium. If this gas heats up as it falls onto the black hole, it will emit a distinctive spectrum of visible radiation. One study sifted through nearly 4 million stellar sources from the Sloan Digital Sky Survey and ended up with forty that had the appropriate optical colors and weak X-ray emission. None has been confirmed as a black hole so the jury is still out on this method.

28   Dark matter is one of the great unsolved problems in cosmology. Star motions in galaxies of all types show that they must be held together by some form of matter that doesn't emit light but adds up to 6 times the sum of all stars. Microlensing surveys were used to show that, at least in the Milky Way, dark matter can't be composed of stellar remnants or sub-stellar objects. Infrared observations additionally rule out rocky objects, all the way from planets to dust grains. The best remaining explanation is a new form of massive, weakly interacting subatomic particle.

29   L. Wyrzykowski, Z. Kostrzewa-Rutkowska, and K. Rybicki, "Microlensing by Single Black Holes in the Galaxy," *Proceedings of the XXXVII Polish Astronomical Society*, 2016. Despite the difficulty, microlensing is an important complement to black hole statistics from binary systems. Black holes in binaries are not found smaller than 6 solar masses, and neutron star masses almost all lie between 1 and 2 solar masses. There seems to be a "gap" in the mass distribution of stellar remnants from 2 to 6 solar masses, potentially challenging current theories of remnant formation. Reassuringly, microlensing shows no such gap.

30   E. A. Poe, "A Descent into the Maelstrom" (1841), in *The Collected Works of Edgar Allan Poe*, edited by T. O. Mabbott (Cambridge, MA: Harvard University Press, 1978).

31   America's venerable Hoover Dam, opened in 1936, generates 25 times less electricity, and doesn't crack the top fifty in the world in terms of power production. The highest peak power production is from the controversial Three Gorges Dam in China, but averaged over a year the Itaipu Dam slightly edges it out.

32   The angular momentum of a particle is mvr, the mass of the particle times its velocity times its distance from the black hole. Kepler's second law shows how angular momentum is conserved in an orbit. When a planet or a comet moves closer to the Sun, it moves faster, so r goes down but v goes up to compensate, and the product is constant.

33   The actual calculation requires general relativity and some numerical approximations. The only more efficient energy production process is matter-antimatter annihilation, which releases mass-energy with 100% efficiency. However, that is a very rare situation in the universe, whereas accretion power is seen coming from all black holes in binary systems. For the full story, see a textbook like J. Frank, A. King, and D. Raine, *Accretion Power in Astrophysics*, 3rd edition, (Cambridge, UK: Cambridge University Press, 2002).

34   The thorny problem was figuring out how angular momentum could be lost to allow matter to fall in. The answer involved turbulence and the role of magnetic fields that thread the accretion disk. The first "standard" accretion disk model, which partially solved the problem, was N. I. Shakura and R. A. Sunyaev, "Black Holes in Binary Systems: Observational Appearance," *Astronomy and Astrophysics* 24 (1973): 337–55. The breakthrough was the realization that magnetic fields can greatly boost angular momentum transport; see S. A. Balbus and J. F. Hawley, "A Powerful Local Shear Instability in Weakly Magnetized Disks: I. Linear Analysis," *Astrophysical Journal* 376 (1991): 214–33. It took the power of modern computers to fully model the situation. 3D magneto-hydrodynamics calculations are among the most challenging in astrophysics.

35   D. Raghavan et al., "A Survey of Stellar Families: Multiplicity of Solar-Type Stars," *Astrophysical Journal Supplement* 190 (2010): 1–42.

36   The imaginary surface defining the region where material is bound to a star in a binary system is called a Roche lobe, after a French astronomer and mathematician of the mid-nineteenth century. Roche lobes are stretched from spheres for isolated stars to teardrop shapes for close binaries. In a detached binary, each star has its own Roche lobe. In a semi-detached binary, the teardrops touch and mass can flow through the point where they meet, the Lagrangian point, named after an Italian astronomer and mathematician of the mid-eighteenth century. In a contact binary, the stars have a common envelope and much of the mass is shared. When the stars have wider separations mass can pass between them if one star is massive and has a wind; a fraction of the gas flowing out in all directions will fall onto the companion.

37   D. Prialnik, "Novae," in *Encyclopedia of Astronomy and Astrophysics*,

edited by P. Murdin (London: Institute of Physics, 2001), 1846–56. About 10 novae are discovered in the Milky Way each year, and they mostly flare up on timescales ranging from 1,000 to 100,000 years. A few spectacular novae flare up within a human lifetime and brighten enough to be visible without a telescope. T Coronae Borealis, or the "Flare Star," flared to become one of the brighter stars in the sky in 1866 and 1946, and RS Ophiuchi has flared enough to be visible with the naked eye five times in the past century, most recently in 2006.

38   This scenario might seem minor and esoteric, but it is central to modern astronomy. Some supernovae (called Type 2) happen when a single massive star dies, but they have luminosities that vary widely. However, when a supernova goes off in a binary system (called Type 1a), it's the result of matter being "spooned onto" a white dwarf in a regulated way, so the luminosity only varies from one system to another by 15%. These supernovae are "standard bombs," so they are also "standard light bulbs" that can be used to measure distance. Since a supernova can be as bright as an entire galaxy, they are visible out to distances of billions of light years. Type 1a supernovae were used to discover the accelerating universe and dark energy in the mid-1990s, work that resulted in a Nobel Prize. See S. Perlmutter, "Supernovae, Dark Energy, and the Accelerating Universe," *Physics Today*, April 2003, 53–60.

39   K. A. Postnov and L. R. Yungelson, "The Evolution of Compact Binary Systems," *Living Reviews in Relativity* 9 (2006): 6–107.

3: SUPERMASSIVE BLACK HOLES

1   The telescope cost him $2,000, equivalent to about $33,000 today. Reber was a one-man shop, laying cement, doing his own metalwork and woodwork, wiring and building the receiver, making the observations, and reducing the data and interpreting it astronomically.

2   As the Earth orbits the Sun, every star rises and sets four minutes earlier each day. That adds up to twenty-four hours over a year, and the whole sky cycles through our night. Star time, or sidereal time, is therefore slightly different from Sun time, or solar time. Jansky used this to show that his radio signal had an extraterrestrial origin, just as Jocelyn Bell did decades later with pulsars.

3   K. Jansky, "Electrical Disturbances Apparently of Extraterrestrial Origin," *Proceedings Institute of Radio Engineers* 21 (1933): 1837. There's a remarkable parallel with the accidental discovery of microwave radiation left over from the big bang three decades later. In 1964, Arno Penzias and Robert Wilson, at Bell Labs, were looking into the feasibility of satellite communications using microwaves. When they tracked down noise sources in their radio receiver they found a weak residual "hiss" that had the same intensity in all directions in the sky. It was radiation from the early universe, cooled and diluted by cosmic expansion. This time, Bell Labs took notice. Penzias and Wilson won the Nobel Prize in Physics in 1978 for their discovery.

4   In honor of his pioneering contribution, the unit of radio radiation strength was named the jansky, so he joined a handful of other electrical pioneers with units named after them: Watt, Volt, Ohm, Hertz, Ampere, and Coulomb. Jansky died in 1950 at the age of forty-four, from Bright's disease that led to kidney failure. He never got to see the rapid growth of the subject he started.

5   Quoted in W. T. Sullivan, ed., *Classics of Radio Astronomy* (Cambridge, UK: Cambridge University Press, 1982).

6   This story is told by John Kraus in *Big Ear* (Delaware, OH: Cygnus-Quasar Books, 1994), and in J. D. Kraus, "Grote Reber, Founder of Radio Astronomy," *Journal of the Royal Astronomical Society of Canada* 82 (1988): 107–13.

7   G. Reber, "Cosmic Static," *Astrophysical Journal* 100 (1944): 279. See also the commentary written for the centennial issue of the journal, K. I. Kellerman, "Grote Reber's Observations of Cosmic Static," *Astrophysical Journal* 525 (1988): 371–72.

8   Kraus, "Grote Reber, Founder of Radio Astronomy."

9   In spectroscopy, the wavelengths of the spectral lines match to elements and give chemical composition, but the nature of the lines indicates the physical situation of the gas. When cooler gas is outside a hotter source of energy, like the outer envelope of a star, absorption lines are seen. That's what von Fraunhofer first saw in the Sun in the early 1800s. When gas is energized so that the electrons are tripped off all the atoms, it leads to a set of emission lines, indicating a very hot source. And when in addition the spectral lines are broad, the large velocity range indicates a violent source of energy to cause the motion of the gas.

10   S. J. Dick, *Discovery and Classification in Astronomy: Controversy and Consensus* (Cambridge, UK: Cambridge University Press, 2013).

11   I was lucky to use the 100-inch telescope on Mount Wilson the year before the Carnegie Institution mothballed it. The growing lights of Los Angeles had made it uncompetitive years before, but it was exciting to use a telescope that had been the world's largest for thirty years, the telescope Edwin Hubble used to show that galaxies were remote from the Milky Way and that the universe was vast and expanding. I remember the row of wooden lockers behind the north pier, one of which had Hubble's name neatly etched on a brass plaque. Perhaps Hubble left his last night lunch inside? Walking on the dome floor, I saw beads of mercury underfoot. The telescope bearings floated on mercury and it leaked; over the years several staff members died from too much contact with it. In Hubble's time, the observers worked for a few hours, stopped for dinner followed by port and cigars, and then resumed their labors. Dinner at Mount Wilson was old-school and formal. The senior astronomer on the mountaintop sat at the head of the table, with other staff astronomers nearby, and the students and postdocs like me at the far end of the table. Dinner was served by a brilliant but tempestuous French chef who had started a number of restaurants around Los Angeles, but each one had failed as he fell out with his patrons and backers. Mount Wilson Observatory was a perfect haven for someone who was creative but had sociopathic

tendencies. The food was sumptuous, but so rich that I found myself hallucinating as the night wore on. To clear my head I walked out onto the catwalk that circled the dome three stories up. Stars twinkled above while the city lights spread out below in a glowing, quilted grid.

12   C. K. Seyfert, "Nuclear Emission in Spiral Galaxies," *Astrophysical Journal* 97 (1943): 28–40.

13   Ryle and Lovell were physicists who clearly saw the power of radio techniques to open up a new window on the universe. They easily moved past the gulf that had separated the engineering and science "cultures" and they each started a major university research group, turning radio astronomy into just another branch of astronomy. The wartime radar expert Robert Dicke started a research group at MIT, but radio astronomy was surprisingly slow to take off in the United States, given that it was the home of Jansky and Reber.

14   Ruby Payne-Scott's contributions have been described in M. Goss, *Making Waves: The Story of Ruby Payne-Scott, Australian Pioneer Radio Astronomer* (Berlin: Springer, 2013). The early story of radio astronomy is excellently told in W. T. Sullivan III, *Cosmic Noise: A History of Early Radio Astronomy* (Cambridge, UK: Cambridge University Press, 2009).

15   The confusion deepened when Ryle and others showed that the radiation from Cygnus A was in fact steady. The variations that had been observed were due to bending of radio waves by clouds of ionized gas in the Earth's upper atmosphere. Ironically, this didn't kill the "radio star" hypothesis because, in optical light, stars twinkle and planets don't. This is because stars are point-like and planets are disk-like, so a planet's twinkling is washed out for an observer on the Earth. By the same logic, if Cygnus A twinkles it must be point-like or at least have a small angular size.

16   B. Lovell, "John Grant Davies (1924–1988)," *Quarterly Journal of the Royal Astronomical Society* 30 (1989): 365–69.

17   The actual formula is $\theta = 1.22\,(\lambda/D)$, where θ is the angular resolution or width of the beam in radians, λ is the wavelength of observation and D is the telescope diameter (measured in the same units).

18   The method is a radio analog of Michelson's interferometer or Young's double slit experiment. Imagine a source directly overhead for two radio dishes. The path length taken by the waves to each dish is the same, so when those waves are combined they add to give a higher amplitude. As the source moves, the path difference changes; when it's half a wavelength the two signals combine and cancel out. So as the source moves, a fringe pattern of high and low signals is created. The width of the interference fringes is set by the separation of the two dishes, which is why the position can be determined so accurately. The radio astronomy group in Australia devised an ingenious version of this idea. They placed an antenna on a sea cliff, facing east. As a radio source rose, the radio radiation would reach the antenna directly at a shallow angle but also traveling a slightly longer path by reflection off the sea's surface. The antenna and its "mirror image" were the two elements of the interferometer.

19   Quoted in the editor's introduction to *Quasi-Stellar Sources and Gravita-*

tional Collapse: Proceedings of the First Texas Symposium on Relativistic Astrophysics, edited by I. Robinson, A. Schild, and E.L. Schucking (Chicago: University of Chicago Press, 1965).

20   Quoted in J. Pfeiffer, *The Changing Universe* (London: Victor Gollancz, 1956).

21   A. Alfven and N. Herlofson, "Cosmic Radiation and Radio Stars," *Physical Review* 78 (1950): 616. Other early papers were G. R. Burbidge, "On Synchrotron Radiation from Messier 87," *Astrophysical Journal* 124 (1956): 416–29, and V. L. Ginzburg and I. S. Syrovaskii, "Synchrotron Radiation," *Annual Reviews of Astronomy and Astrophysics* 3 (1965): 297–350.

22   Substantial technical problems had to be overcome to match strong radio sources to optical counterparts. Different radio surveys didn't always agree on the strength or even the existence of a particular source. Radio sources have angular sizes that range from tens of arc minutes down to a few arc seconds, and what's seen with an interferometer depends on the number of elements in the array and their spacing, as well as the frequency of observation. Also, the number of radio sources in any particular area of sky increases quite rapidly with decreasing radio flux. That means there can be multiple sources near the limit of detection that conspire to look like a single, stronger source. This is called the "confusion limit" of a survey.

23   C. Hazard, M. B. Mackey, and A. J. Shimmins, "Investigation of the Radio Source 3C 273 by the Method of Lunar Occultations," *Nature* 197 (1963): 1037–39; M. Schmidt, "3C 273: A Star-like Object with Large Redshift," *Nature* 197 (1963): 1040; J. B. Oke, "Absolute Energy Distribution in the Optical Spectrum of 3C 273," *Nature* 1987 (1963): 1040–41; and J. L. Greenstein and T. A. Matthews, "Redshift of the Unusual Radio Source: 3C 48," *Nature* 197 (1963): 1041–42. For a modern summary of the chronology, see C. Hazard, D. Jauncey, W. M. Goss, and D. Herald, "The Sequence of Events that led to the 1963 Publications in Nature of 3C 273, the first Quasar and the first Extragalactic Radio Jet," in *Proceedings of IAU Symposium 313*, edited by F. Massaro et al. (Dordrecht: Kluwer, 2014).

24   Interview with Maarten Schmidt on the fiftieth anniversary of his discovery, http://www.space.com/20244-quasar-mystery-discoverer-interview.html.

25   In fact, Australian radio astronomer John Bolton and American astronomer Allan Sandage each had a spectrum of 3C 48 in 1960, and both narrowly missed making the first quasar discovery three years before Schmidt.

26   Cosmological redshift is physically distinct from a Doppler shift. Doppler shift occurs when a wave travels in a medium and the source of the wave is moving with respect to the observer. The common example is a siren that rises in pitch when a police car is approaching and falls in pitch when the car is receding. Cosmological redshift does not require a medium, because the change in wavelength is caused by the expansion of space-time everywhere in the universe.

27   Cosmology broadens the Copernican principle, the idea that we don't occupy a privileged location in the Solar System, to the whole universe. It's a fundamental assumption of modern cosmology that hasn't been violated by any observation so far. The galaxies near the Milky Way don't appear to be any different, or distributed differently, from galaxies in remote parts of the universe.

28   Hubble's law is $v = H_0 D$, where v is the recession velocity, D is the distance, and the constant of proportionality is the Hubble constant, or the current expansion rate of the universe. The low redshift approximation in terms of the recession velocity and the speed of light is $z = v/c$. The correct relativistic formula is $z = \sqrt{(1 + v/c)/(1 - v/c)}$.

29   M. Schmidt, "Large Redshifts of Five Quasi-Stellar Sources," *Astrophysical Journal* 141 (1965): 1295–1300.

30   F. Zwicky and M. A. Zwicky, *Catalogue of Selected Compact Galaxies and of Post-Eruptive Galaxies* (Guemligen, Switzerland: Zwicky, 1971). The paper that raised his ire was A. Sandage, "The Existence of a Major New Constituent of the Universe: The Quasi-Stellar Galaxies," *Astrophysical Journal* 141 (1965): 1560–68. The episode is recounted in K.I. Kellerman, "The Discovery of Quasars and its Aftermath" *Journal of Astronomical History and Heritage* 17 (2014): 267–82.

31   The next phase of giant telescope building is as fiercely competitive as the last phase. Each of the 20-meter or larger telescopes planned will cost a billion dollars or more. The Giant Magellan Telescope is on the inside track, with five of the seven mirrors already cast at the University of Arizona, and a mountaintop scraped off and construction started in Chile. The Caltech project to build a 30-meter telescope was stalled due to opposition on Mauna Kea from native Hawaiian activists, but is now back on track. The 39-meter telescope of the European Southern Observatory will also go to Chile, and it is well funded by international agreements among the mostly European partners. The dark horse in the race is China, which may leapfrog over the 8–10-meter class to build another giant telescope on the Tibetan plateau.

32   The Seyfert calculation was presented in L. Woltjer, "Emission Nuclei in Galaxies," *Astrophysical Journal* 130 (1959): 38–44. The energy calculation for radio galaxies was presented in G. Burbidge, "Estimates of the Total Energy and Magnetic Field in the Non-Thermal Radio Sources," *Astrophysical Journal* 129 (1959): 849–52.

33   V. Ambartsumian, "On the Evolution of Galaxies," in *The Structure and Evolution of the Universe*, edited by R. Stoops (Brussels: Coudenberg, 1958), 241–74.

34   E. Salpeter, "Accretion of Interstellar Matter by Massive Objects," *Astrophysical Journal* 140 (1964): 796–800; Ya. B. Zel-dovich, "On the Power Source for Quasars," *Soviet Physics Doklady* 9 (1964): 195–205.

35   The main proponents of noncosmological redshifts in the 1960s and through the 1970s were Halton Arp and Bill Tifft on the observational side and Fred Hoyle and Geoff Burbidge on the theoretical side. The quasar redshift "controversy" was the subject of heated debate at conferences, with little agreement between the two sides. The debate was largely settled in favor of the cosmological interpretation by the 1980s, but even now there are some researchers who claim quasars are not at the distances indicated by their redshifts. The observational arguments can be seen in H. C. Arp, "Quasar Redshifts," *Science* 152 (1966): 1583, and the theoretical argument can be seen in G. Burbidge and F. Hoyle, "The Problem of the Quasi-Stellar Objects," *Scientific American* 215 (1966): 40–52.

36   All the radio emission is from electrons emitting synchrotron radiation in a hot but diffuse plasma. The energy transport must be very efficient to reach so far beyond the galaxy. The lobes present the places where relativistic particles "hit" the diffuse intergalactic medium, often creating hot spots of enhanced emission. The hot plasma is threaded by magnetic fields which means the radio emission has linear polarization.

37   D. S. De Young, *The Physics of Extragalactic Radio Sources* (Chicago: University of Chicago Press, 2002).

38   The discovery papers were A. R. Whitney et al., "Quasars Revisited: Rapid Time Variations Observed Via Very Long Baseline Interferometry," *Science* 173 (1971): 225–30, and M. H. Cohen et al., "The Small Scale Structure of Radio Galaxies and Quasi-Stellar Sources at 3.8 Centimeters," *Astrophysical Journal* 170 (1971): 207–17. Apparent superluminal motion had been predicted, based on theoretical arguments, five years earlier, in M. J. Rees, "Appearance of Relativistically Expanding Radio Sources," *Nature* 211 (1966): 468–70.

39   A.-K. Baczko et al., "A Highly Magnetized Twin-Jet Base Pinpoints a Supermassive Black Hole," *Astronomy and Astrophysics* 593 (2016): A47–58.

40   The ionized regions around stars also show strong emission lines, but the lines in Seyfert spectra require a lot of ultraviolet radiation to be excited, more than can be generated by young stars. Spectroscopy was used to classify Seyferts as either Type 1, with very broad emission lines indicating gas motions up to 5% of the speed of light, or Type 2, with narrower emission lines. Seyferts 1s are generally more luminous than the Seyfert 2s, and there's even an intermediate class of Seyfert 1.5 galaxies, where the emission lines have weak broad wings superimposed on strong narrow cores. Astronomers also found a category of galaxies with low excitation nuclear emission lines, or LINERs, which are more active than a normal galaxy but less active than a Seyfert galaxy. Yes, the taxonomy of active galaxies is complex and confusing.

41   This type of observation of quasar "host galaxies" in the 1990s helped lay to rest the claim that quasar redshifts were noncosmological. There was a continuum of active nuclei ranging from mild ones nearby to very remote and luminous ones, where the data was consistent with them living in a galaxy at the distance indicated by the redshift in an expanding universe. Meanwhile, some of the evidence for noncosmological redshifts evaporated. There was no excess of redshifts at particular values, the distribution was smooth, and apparent associations of high redshift quasar with low redshift galaxies were shown to be coincidences and not indicative of a physical connection.

42   R. D. Blandford and M. J. Rees, "Some Comments on the Radiation Mechanism in Lacertids," in *Pittsburgh Conference on BL Lac Objects*, edited by A.M. Wolfe (Pittsburgh: University of Pittsburgh, 1978).

43   C. S. Bowyer et al., "Detection of X-Ray Emission from 3C 273 and NGC 5128," *Astrophysical Journal* 161 (1970): L1–L7.

44   The first sensitive study of quasar X-ray emission was H. Tananbaum et al., "X-Ray Studies of Quasars with the Einstein Observatory," *Astrophysical Journal* 234 (1979): L9–13. Greg Shields was the first to suggest that quasar

UV emission was due to an accretion disk, in G. A. Shields, "Thermal Emission from Accretion Disks in Quasars," *Nature* 272 (1978): 706–08. Matt Malkan was the first to derive detailed accretion disk models, in M. A. Malkan, "The Ultraviolet Excess of Luminous Quasars: II. Evidence for Massive Accretion Disks," *Astrophysical Journal* 268 (1983): 582–90.

45   D. B. Sanders et al., "Continuum Energy Distribution of Quasars – Shapes and Origins," *Astrophysical Journal* 347 (1979): 29–51.

46   IceCube Collaboration, "Neutrino emission from the Direction of the Blazar TXS 0506+056 Prior to the IceCube-170922A Alert," *Science* 361 (2018), 147–51.

47   I did my PhD on blazars, and I was drawn to them because they offered the clearest view of the maelstrom. At each observing run, I had a "hot list" of several dozen targets where monitoring with small telescopes had shown signs of unusual activity. Sometimes the target was a bust, the light trace as flat as a millpond. Other times, the central black hole was gorging on gas and stars and cranking out high-energy radiation and electrons traveling at 99.999% of the speed of light. Like Poe's fictional narrator, I was drawn by the terrible beauty of a deep and unforgiving gravitation pit.

48   M. A. Orr and I. W. A. Browne, "Relativistic Beaming and Quasar Statistics," *Monthly Notices of the Royal Astronomical Society* 200 (1982): 1067–80. A relativistic jet oriented close to the line of sight can easily have its flux boosted by a factor of 1,000. The counter-jet is moving rapidly away from the observer so is de-amplified; the result for the observer is a one-sided jet. Extended radio emission is not part of a relativistic flow so its flux is unaffected.

49   The progress of this idea can be followed in two review articles separated by more than twenty years: R. R. J. Antonucci, "Unified Models for Active Galactic Nuclei and Quasars," *Annual Reviews of Astronomy and Astrophysics* 31 (1993): 473–521; and H. Netzer, "Revisiting the Unified Model of Active Galactic Nuclei," *Annual Reviews of Astronomy and Astrophysics* 53 (2015): 365–408.

4: GRAVITATIONAL ENGINES

1   The most famous representation of this myth is Tintoretto's *The Origin of the Milky Way* (1575), in the National Gallery in London. Most people in Western countries live in cities and suburbs and their view of the Milky Way is obscured by light pollution. When I survey millennials in large classes I teach at the University of Arizona, typically only 10% have ever seen the Milky Way.

2   Z. M. Malkin, "Analysis of Determinations of the Distance between the Sun and the Galactic Center," *Astronomy Reports* 57 (2013): 128–33.

3   W. M. Goss, R. L. Brown, and K. Y. Lo, "The Discovery of Sgr A°," in "Proceedings of the Galactic Center Workshop – The Central 300 Parsecs of the Milky Way," *Astronomische Nachrichen*, supplementary issue 1 (2003): 497–504.

4   M. J. Rees, "Black Holes," *Observatory* 94 (1974): 168–79.

5   Infrared detectors were often developed for military applications like nighttime battlefield imaging and heat-tracking of missiles, which slowed their adoption in

the civilian and research sectors. Also, infrared imaging must deal with thermal background radiation that is millions of times higher than the optical radiation from a dark night sky. For the overall history of the subject, see G. H. Rieke, "History of Infrared Telescopes and Astronomy," *Experimental Astronomy* 125 (2009): 125–41. For a history of detector development, see A. Rogalski, "History of Infrared Detectors," *Opto-Electronics Review* 20 (2012): 279–308. Optical astronomy took its big leap forward in the late 1970s when charged coupled devices (CCDs) migrated from research labs to astronomical use.

6   The crowding of images in a dense region of stars, or the smooth distribution of light in a galaxy image, is the result of images that are far bigger than the stars themselves. Starlight is blurred by a particular amount when it passes through the Earth's atmosphere, regardless of the size of the source of light. Stars in our part of the Milky Way are widely separated and almost never collide; the distances between them are millions of times larger than their sizes. Even in the central part of the Milky Way, the distances between stars are tens of thousands of times larger than their sizes and they almost never collide.

7   From the German group, A. Eckart and R Genzel, "Observations of Stellar Proper Motions Near the Galactic Centre," *Nature* 383 (1996): 415–17, and A. Eckart and R. Genzel, "Stellar Proper Motions in the Central 0.1 pc of the Galaxy," *Monthly Notices of the Royal Astronomical Society* 28 (1997): 576–98; and from the American group, A. M. Ghez, B. L. Klein, M. Morris, and E. E. Becklin, "High Proper Motion Stars in the Vicinity of Sagittarius A°: Evidence for a Supermassive Black Hole at the Center of our Galaxy," *Astrophysical Journal* 509 (1998): 678–86.

8   Quoted in http://www.pbs.org/wgbh/nova/space/andrea-ghez.html.

9   Reinhard Genzel explains why it's so important to have a massive black hole on our doorstep, thousands of times closer than any other active galaxy or quasar: "The center of our galaxy is a unique laboratory where we can study the fundamental processes of strong gravity, stellar dynamics, and star formation that are of great relevance to all other galactic nuclei, and with a level of detail that will never be possible beyond our Galaxy." Quoted in http://www.universetoday.com/22104/beyond-any-reasonable-doubt-a-supermassive-black-hole-lives-in-centre-of-our-galaxy/.

10   Andrea Ghez left behind the tentativeness of a young researcher long ago. She's a superstar and a role model for young women in astronomy. Ghez was elected to the National Academy of Sciences before she turned forty, and in 2008 was awarded a MacArthur Fellowship, colloquially known as the "genius prize." She's unaffected by all the celebrity and enjoys talking about having as much fun as when she solved puzzles as a young child: "Research is a wonderful career, because once you've started to work on one question, what you find is not only the answer to the first question but new puzzles. I think that's what keeps me going, there are always open questions, new puzzles."

11   F. Roddier, *Adaptive Optics in Astronomy* (Cambridge, UK: Cambridge University Press, 2004).

12   A. M. Ghez et al., "Measuring Distance and Properties of the Milky Way's

Supermassive Black Hole with Stellar Orbits," *Astrophysical Journal* 689 (2008): 1044–62; and S. Gillesen et al., "Monitoring Stellar Orbits Around the Massive Black Hole in the Galactic Center," *Astrophysical Journal* 692 (2009): 1075–1109.

13   S. Gillesen et al., "A Gas Cloud on its Way Towards the Supermassive Black Hole in the Galactic Centre," *Nature* 481 (2012): 51–54.

14   S. Doeleman et al., "Event-Horizon Scale Structure in the Supermassive Black Hole Candidate at the Galactic Centre," *Nature* 455 (2008): 78–80.

15   A. Boehle et al., "An Improved Distance and Mass Estimate for Sgr A° from Multistar Orbit Analysis," *Astrophysical Journal*, 830 (2016): 17-40.

16   M. Schmidt, "The Local Space Density of Quasars and Active Nuclei," *Physica Scripta* 17 (1978): 135–36.

17   D. Lynden-Bell, "Galactic Nuclei as Collapsed Old Quasars," *Nature* 223 (1969): 690–94.

18   The formula for the gravitational influence radius is $R_g = GM/v^2$, where M is the black hole mass in solar masses and v is the dispersion or spread in velocities of stars within that radius, caused by both the black hole and the stars themselves. Based on the observed scaling relations between black hole mass and stellar velocity dispersion, this becomes $R_g \approx 35 \, (M/10^9)^{1/2}$ parsecs.

19   Combining the formula for the gravitational radius $R_g = GM/v^2$ with the formula for the Schwarzschild radius $R_S = GM/c^2$ leads to $R_g/R_S = (c/v)^2$, which is about 10^6 for a massive galaxy, where $v = 200$–300 kilometers per second.

20   R. F. Zimmerman, *The Universe in a Mirror: The Saga of the Hubble Space Telescope and the Visionaries Who Built It* (Princeton: Princeton University Press, 2010).

21   I used Hubble back when it was first launched and many times since. "Used" is a euphemism, since even experienced astronomers don't get to move the telescope around and watch as faint galaxies swim into view. With a price tag of $8 billion, it's far too valuable to risk a malfunction caused by a careless user. After getting allocated orbits in a fiercely competitive review process, astronomers submit their target lists and an AI scheduling algorithm interleaves them to minimize energy use, instrument changes, and time spent slewing the telescope. A few weeks later the reduced data is available to download from a secure website. Not very romantic, alas.

22   Inevitably, there are complications and subtleties. Galaxies are three-dimensional objects, so three-dimensional space motions project down to two dimensions on the plane of the sky, and a spectrograph slit can only sample a one-dimensional map of the dispersion in velocities. As a result, the data has to be modeled and assumptions are made in the analysis. Different slit orientations can be used to get close to a two-dimensional velocity map, but that requires a lot of coveted telescope time for each galaxy.

23   L. Ferrarese and D. Merritt, "Supermassive Black Holes," *Physics World* 15 (2002): 41–46; and L. Ferrarese and H. Ford, "Supermassive Black Holes in Galactic Nuclei: Past, Present, and Future," *Space Science Reviews* 116 (2004): 523–624.

24   R. Bender et al., "HST STIS Spectroscopy of the Triple Nucleus of M31: Two Nested Disks in Keplerian Motion around a Supermassive Black Hole," *Astrophysical Journal* 631 (2005): 280–300.

25   R. P. van der Marel, P. T. de Zeeuw, H.-W. Rix, and G. D. Quinlan, "A Massive Black Hole at the Center of the Quiescent Galaxy M32," *Nature* 385 (1997): 610–12.

26   K. Gebhardt and J. Thomas, "The Black Hole Mass, Stellar Mass-to-Light Ratio, and Dark Halo in M87," *Astrophysical Journal* 700 (2009): 1690–1701.

27   M. C. Begelman, R. D. Brandford, and M. J. Rees, "Theory of Extragalactic Radio Sources," *Reviews of Modern Physics* 56 (1984): 255–351.

28   R. D. Blandford, H. Netzer, and L. Woltjer, *Active Galactic Nuclei* (Berlin: Springer, 1990).

29   M. C. Begelman and M. J. Rees, "The Fate of Dense Stellar Systems," *Monthly Notices of the Royal Astronomical Society* 185 (1978): 847–60; and M. C. Begelman and M. J. Rees, *Gravity's Fatal Attraction: Black Holes in the Universe* (Cambridge, UK: Cambridge University Press, 2009).

30   P. Khare, "Quasar Absorption Lines: an Overview," *Bulletin of the Astronomical Society of India* 41 (2013): 41–60.

31   W. L. W. Sargent, "Quasar Absorption Lines and the Intergalactic Medium," *Physica Scripta* 21 (1980): 753–58.

32   D. H. Weinberg, R. Dave, N. Katz, and J. Kollmeier, "The Lyman-Alpha Forest as a Cosmological Tool," in *The Emergence of Cosmic Structure*, AIP Conference Series 666, edited by S. Holt and C. Reynolds, 2003, 157–69.

33   In the theory of lensing, there are always an odd number of images, some magnified and some demagnified. The most common lensing geometry creates a pair of magnified images and a demagnified image that's usually too faint to detect, so a pair of images is seen. If the mass distribution of the lensing object is complex, higher image multiplicities are possible, so astronomers have seen 4-image, 6-image, and even 10-image lensed quasars. For a summary of the phenomenon, see T. Sauer, "A Brief History of Gravitational Lensing," Einstein Online, Volume 4, 2010, http://www.einstein-online.info/spotlights/grav_lensing_history.

34   The U.K. Schmidt is well described by former staff member Fred Watson in *Stargazer: Life and Times of the Telescope* (London: Allen and Unwin, 2004). See also a summary at https://www.aao.gov.au/about-us/uk-schmidt-telescope-history.

35   M. Miyoshi et al., "Evidence for a Black Hole from High Rotation Velocities in a Sub-Parsec Region of NGC4258," *Nature* 373 (1995): 127–29.

36   A. J. Baarth et al., "Towards Precision Black Hole Masses with ALMA: NGC 1332 as a Case Study in Molecular Disk Dynamics," *Astrophysical Journal* 823 (2016): 51–73.

37   B. M. Peterson, "The Broad Line Region in Active Galactic Nuclei," *Lecture Notes in Physics* vol. 693 (Berlin: Springer, 2006), 77–100.

38   There are many details and complications involved in making a reliable mass estimate. The fast-moving gas that gives the emission lines is in clouds rather

than being smoothly distributed, and clouds of different densities and distance from the black hole emit different emission lines. The geometry of the gas affects the time delay signal. For example, a ringlike geometry of gas has a constant time delay surface that's a parabola. A more complicated 3D geometry for the gas makes the analysis challenging. Uneven sampling of the variability due to the vagaries of weather and telescope scheduling create more headaches. Up to 100 astronomers may be involved in one of these intense reverberation mapping campaigns, all to bag a handful of black hole masses.

39   M. C. Bentz et al., "NGC 5548 in a Low-Luminosity State: Implications for the Broad-Line Region," *Astrophysical Journal* 662 (2007): 205–12.

40   B. M. Peterson and K. Horne, "Reverberation Mapping of Active Galactic Nuclei," in *Planets to Cosmology: Essential Science in the Final Years of the Hubble Space Telescope,* edited by M. Livio and S. Casertano (Cambridge, UK: Cambridge University Press, 2004).

41   The methods are summarized in B. M. Peterson, "Measuring the Masses of Supermassive Black Holes," *Space Science Review* 183 (2014): 253–75. A large quantity of data is presented in A. Refiee and P. B. Hall, "Supermassive Black Hole Mass Estimates Using Sloan Digital Sky Survey Quasar Spectra at $0.7 < z < 2$," *Astrophysical Journal Supplements* 194 (2011): 42–58.

42   To put this number in perspective, world energy consumption is about 20 terawatts, which is a million billion billion times (10^{26}) less power than is pumped out by a quasar.

43   J. Updike, "Ode to Entropy," in *Facing Nature* (New York: Knopf, 1985).

44   The fundamental physical distinction is between a thermal process and a nonthermal process. In a thermal process, the physical system is in equilibrium and has a characteristic temperature. In this case, it emits blackbody radiation over a range of wavelengths but with a well-defined peak where the wavelength of the peak emission is inversely proportional to temperature (Wein's law). In a nonthermal process, the physical system is out of equilibrium and has no characteristic temperature. Radiation is emitted over a very broad range of wavelengths, typically with a power law energy distribution. Synchrotron radiation is an example of nonthermal radiation, as in the radio emission from active galaxies and quasars.

45   A. Prieto, "Spectral Energy Distribution Template of Redshift-Zero AGN and the Comparison with that of Quasars," in *Astronomy at High Angular Resolution,* Journal of Physics Conference Series, vol. 372 (London: Institute of Physics, 2012), 1–5.

46   X. Barcons, *The X-Ray Background* (Cambridge, UK: Cambridge University Press, 1992).

47   A. Moretti et al., "Spectrum of the Unresolved Cosmic X-Ray Background: What is Unresolved 50 Years after its Discovery?" *Astronomy and Astrophysics* 548 (2012): 87–99.

48   Some of the most common misconceptions are dealt with neatly by Phil Plait, a.k.a. the Bad Astronomer, on his blog for *Discover,* http://blogs.discovermagazine.com/badastronomy/2008/10/30/ten-things-you-dont-know-about-black-holes/ - .WEoS2horJdg.

5: THE LIVES OF BLACK HOLES

1   B. J. Carr and S. Hawking, "Black Holes in the Early Universe," *Monthly Notices of the Royal Astronomical Society* 168 (1974): 399–415.

2   The Planck time is part of a system of units often used in particle physics and cosmology, where measurement is defined entirely in terms of fundamental constants and not human-derived constructs. By convention, physical constants take values of 1 when calculating in Planck units. Planck units describe a situation where the standard quantum theory and general relativity cannot be reconciled, and a quantum gravity theory is needed. This occurs at the Planck energy of 10^{19} GeV.

3   The alternative to hypothesizing dark matter is to say that Newton's law of gravity is wrong. If the gravity force didn't depend exactly on the inverse square of the distance, it would be possible to explain away the need for dark matter. But the price paid would be high. Newton's law of gravity is preeminent in explaining weak gravity in the Solar System and beyond, and altering the force law destroys the symmetry and elegance of the theory. Various alternative gravity theories have been explored, but none passes the high bar cleared by Newton's theory. Astronomers have accepted that dark matter is a major component of the universe, and major efforts are devoted to figuring out its physical nature.

4   P. Pani and A. Loeb, "Exclusion of the Remaining Mass Window for Primordial Black Holes as the Dominant Constituent of Dark Matter," *Journal of Cosmology and Astroparticle Physics*, issue 6 (2014): 26.

5   S. Singh, *Big Bang: The Origin of the Universe* (New York: Harper Perennial, 2005).

6   J. Miralda-Escude, "The Dark Age of the Universe," *Science* 300 (2003): 1904–09.

7   A. Loeb, "The Habitable Epoch of the Early Universe," *International Journal of Astrobiology* 13 (2014): 337–39.

8   While astronomers don't know the physical nature of dark matter, there's a huge body of evidence saying that invisible mass exists throughout the universe and acts to hold galaxies together. Simulations of structure formation don't generate anything like the real universe unless dark matter is an ingredient. The requirement is for "cold dark matter," where cold means the particle was moving at nonrelativistic speeds when stable atoms formed (otherwise, structures would be erased). The foundational paper is G. R. Blumenthal et al., "Formation of Galaxies and Large-Scale Structures with Cold Dark Matter," *Nature* 31 (1984): 517–25.

9   V. Bromm et al., "Formation of the First Stars and Galaxies," *Nature* 459 (2009): 49–54; and A. Loeb, *How Did the First Stars and Galaxies Form* (Princeton: Princeton University Press, 2010).

10   D. G. York et al., "The Sloan Digital Sky Survey: Technical Summary," *Astronomical Journal* 120 (2000): 1579–87.

11   E. Chaffau et al., "A Primordial Star in the Heart of the Lion," *Astronomy and Astrophysics* 542 (2012): 51–64.

12   G. Schilling, *Flash! The Hunt for the Biggest Explosions in the Universe* (Cambridge, UK: Cambridge University Press, 2002).

13   R. W. Klebasadel, I. B. Strong, and R. A. Olsen, "Observations of Gamma Ray Bursts of Cosmic Origin," *Astrophysical Journal Letters* 182 (1973): L85–89.

14   J. S. Bloom et al., "Observations of the Naked Eye GRB 080319B: Implications of Nature's Brightest Explosion," *Astrophysical Journal* 691 (2009): 723–37.

15   N. Tanvir et al., "A Gamma Ray Burst at a Redshift of z = 8.2," *Nature* 461 (2009): 1254–57.

16   Hunting gamma ray bursts involves a network of telescopes, so that the largest telescope with clear weather can look for an optical counterpart. It's exciting work, but the yield is small. Out of over 5,000 gamma ray bursts in the past fifteen years, less than twenty have been observed quickly enough or had a bright enough optical counterpart for a redshift to be measured.

17   N. Gehrels and P. Meszaros, "Gamma Rays Bursts," *Science* 337 (2012): 932–36.

18   S. Dong et al., "ASASSN-15lh: A Highly Super-Luminous Supernova," *Science* 351 (2016): 257–60.

19   A. L. Melott et al., "Did a Gamma Ray Burst Initiate the Late Ordovician Mass Extinction?" *International Journal of Astrobiology* 3 (2004): 55–61. Also, B. C. Thomas et al., "Gamma Ray Bursts and the Earth: Exploration of Atmospheric, Biological, Climatic, and Biogeochemical Effects," *Astrophysical Journal* 634 (2005): 509–33.

20   V. V. Hambaryan and R. Neuhauser, "A Galactic Short Gamma Ray Burst as Cause for the Carbon-14 Peak in AD 774/775," *Monthly Notices of the Royal Astronomical Society* 430 (2013): 32–36.

21   The physical nature of ULX sources is controversial. They might be accreting black holes, but some of them may be accreting neutron stars. Also, theorists have proposed ways that black holes can be "force fed" and so radiate beyond the Eddington limit, which in turn would mean the black hole need not be so massive. Evidence that a ULX in the nearby galaxy M82 is an intermediate-mass black hole is given in D. R. Pasham, T. E. Strohmayer, and R. F. Mushotzky, "A 400-Solar-Mass Black Hole in the Galaxy M82," *Nature* 513 (2014): 74–76.

22   D. H. Clark, *The Quest for SS433* (New York: Viking, 1985).

23   I. F. Mirabel and R. F. Rodriguez, "Microquasars in our Galaxy," *Nature* 392 (1998): 673–76.

24   L. Ferrarese and D. Merritt, "A Fundamental Relation Between Supermassive Black Holes and Their Host Galaxies," *Astrophysical Journal Letters* 539 (2000): L9–12; and K. Gebhardt et al., "A Relationship Between Nuclear Black Hole Mass and Galaxy Velocity Dispersion," *Astrophysical Journal Letters* 539 (2000): L13–16. This relation was extended to lower-mass dwarf galaxies, both active and inactive, by Jenny Greene and collaborators.

25   T. Oka et al., "Signature of an Intermediate-Mass Black Hole in the Central Molecular Zone in our Galaxy," *Astrophysical Journal Letters* 816 (2015): L7–12.

26   R. Geroch, *General Relativity from A to B* (Chicago: University of Chicago Press, 1981). An excellent set of introductory-level articles can be found at http://www.einstein-online.info/.

27   For a site that tracks the world's fastest 500 computers, and other trends in processor power and computation, see https://www.top500.org/.

28   M. W. Choptuik, "The Binary Black Hole Grand Challenge Project," in *Computational Astrophysics*, edited by D.A. Clarke and M.J. West, ASP Conference Series #123, 1997, 305. This was followed by J. Baker, M. Campanelli, and C. O. Lousto, "The Lazarus Project: A Pragmatic Approach to Binary Black Hole Evolutions," *Physical Review D* 65 (2002): 044001–16.

29   J. Healy et al., "Superkicks in Hyperbolic Encounters of Binary Black Holes," *Physical Review Letters* 102 (2009): 041101–04.

30   The following paper is not for the faint of heart: R. Gold et al., "Accretion Disks Around Binary Black Holes of Unequal Mass: General Relativistic Magnetohydrodynamic Simulations of Postdecoupling and Merger," *Physical Review D* 90 (2014): 104031–45.

31   I saw another side of Simon White while he was my colleague on the astronomy faculty at the University of Arizona. Simon was the go-to guy on any topic in cosmology; his expertise was both wide and deep. He had the skill of transferring to you his physical intuition. I often left a conversation with him thinking I was smarter than I was. He retained several quirks that marked him out as a native Brit. The most striking was on display one evening when I went to his house for a potluck dinner. After the food was finished, table and chairs were pushed to the side, and Simon led out a group of men wearing bells on their shins, knotted handkerchiefs on their heads, and carrying sticks. What followed was Morris dancing, an unbroken tradition since Shakespearean times in the small town in Kent where Simon was born. I grew up in Britain but never imagined I'd see Morris dancing in the Sonoran Desert.

32   E. Bertschinger, "Simulations of Structure Formation in the Universe," *Annual Review of Astronomy and Astrophysics* 36 (1998): 599–654.

33   These methods reduce the computational load for N particles from N^2 to N logN. So for a million particles, there are 6 million calculations, and for 10 billion particles there are 10 million calculations.

34   J. J. Monaghan, "Smoothed Particle Hydrodynamics," *Annual Reviews of Astronomy and Astrophysics* 30 (2002): 543–74.

35   See interview with Simon White at http://www.drillingsraum.com/simon-white/simon-white-1.html.

36   V. Springel et al., "Simulations of the Formation, Evolution, and Clustering of Galaxies and Quasars," *Nature* 435 (2005): 629–36.

37   M. Vogelsberger et al., "Properties of Galaxies Reproduced by a Hydrodynamical Simulation," *Nature* 509 (2014): 177–82.

38   See interview with Simon White at http://www.drillingsraum.com/simon-white/simon-white-4.html.

39   The only galaxies visible to the naked eye are the spiral Andromeda or M31 in

the north, and the Large and Small Magellanic Clouds, two dwarf galaxies visible in the south. With the large fraction of people living in cities and suburbs and being unfamiliar with the night sky, most people have never seen another galaxy.

40   E. Bañados et al., "An 800-Million-Solar-Mass Black Hole in a Significantly Neutral Universe at a Redshift of 7.5," *Nature*, December 6, 2017, doi:10.1038/nature25180. The previous record-holder was D. J. Mortlock et al., "A Luminous Quasar at a Redshift of z = 7.085," *Nature* 474 (2011): 616–19

41   J. L. Johnson et al., "Supermassive Seeds for Supermassive Black Holes," *Astrophysical Journal* 771 (2013): 116–25.

42   A. C. Fabian, "Observational Evidence of AGN Feedback," *Annual Review of Astronomy and Astrophysics* 50 (2012): 455–89.

43   The phenomenon where small galaxies form before large galaxies, while large black holes form before smaller black holes, is called cosmic downsizing. The reason is that the accepted view of galaxy evolution is that small galaxies form first and merge to form larger galaxies. Black holes take a different route, the largest ones growing quickly and the more abundant smaller ones growing slowly and later. Downsizing refers to the tendency for most black holes to grow slowly and stay relatively small. For a review from the simulation standpoint, see P. F. Hopkins et al., "A Unified, Merger-Driven Model of the Origin of Starbursts, Quasars, the Cosmic X-Ray Background, Supermassive Black Holes, and Galaxy Spheroids," *Astrophysical Journal Supplement* 163 (2006): 1–49. For the observational perspective, see M. Volonteri, "The Formation and Evolution of Massive Black Holes," *Science* 337 (2012): 544–47.

44   C. H. Lineweaver and T. M. Davis, "Misconceptions About the Big Bang," *Scientific American*, March 2005, 36–45.

45   See Ned Wright's cosmology FAQ at http://www.astro.ucla.edu/~wright/cosmology_faq.html.

46   N. J. Poplawski, "Cosmology with Torsion: An Alternative to Cosmic Inflation," *Physics Letters B* 694 (2010): 181–85.

47   R. Pourhasan, N. Afshordi, and R. B. Mann, "Out of the White Hole: A Holographic Origin for the Big Bang," *Journal of Cosmology and Astroparticle Physics*, issue 4 (2014): 5–22. A popular version, and the source of the quote, is N. Afshordi, R. B. Mann, and R. Pourhasan, "The Black Hole at the Beginning of Time," *Scientific American*, August 2014, 37–43.

48   J. Tanaka, T. Yamamura, and J. Kanzaki, "Study of Black Holes with the Atlas Detector at the LHC," *European Physical Journal C* 41 (2005): 19–33.

49   CMS Collaboration, "Search for Microscopic Black Hole Signatures at the Large Hadron Collider," *Physics Letters B* 697 (2011): 434–53.

50   B. Koch, M. Bleicher, and H. Stocker, "Exclusion of Black Hole Disaster Scenarios at the LHC," *Physics Letters B* 672 (2009): 71–76.

51   See Ethan Siegel's blog at http://www.forbes.com/sites/startswithabang/2016/03/11/could-the-lhc-make-an-earth-killing-black-hole/#6b465d245837.

52   L. Crane and S. Westmoreland, "Are Black Hole Starships Possible?," 2009, https://arxiv.org/abs/0908.1803.

6: BLACK HOLES AS TESTS OF GRAVITY

1   J. Lequeux, *Le Verrier: Magnificent and Detestable Astronomer* (New York: Springer, 2013). Le Verrier beat English astronomer James Couch Adams to the discovery by only a few days, although Adams completed his work earlier. Le Verrier was so unpopular as director of the Paris Observatory that he was driven out of the job, but he regained the position after his successor accidentally drowned. A contemporary said of him, "I do not know whether M. Le Verrier is actually the most detestable man in France, but I am quite certain he is the most detested." In a fascinating historical twist, Galileo missed the discovery of Neptune over 200 years earlier. In 1613, he noticed a bright object close to Jupiter, but assumed it was a star. He even noticed the object move slightly. However, the following nights were cloudy so Galileo missed making the observations that would have made it clear that he was seeing a planet.

2   R. Baum and W. Sheehan, *In Search of Planet Vulcan: The Ghost in Newton's Clockwork Machine* (New York: Plenum Press, 1997).

3   W. Isaacson, *Einstein: His Life and Universe* (New York: Simon & Schuster, 2007).

4   G. Musser, *Spooky Action at a Distance: The Phenomenon That Reimagines Space and Time—And What It Means for Black Holes, the Big Bang, and Theories of Everything* (New York: Farrar, Straus and Giroux, 2015). See also the more technical but masterful T. Maudlin, *Quantum Non-Locality and Relativity: Metaphysical Intimations of Modern Physics* (Oxford: Wiley–Blackwell, 2011).

5   R. Oerter, *The Theory of Almost Everything: The Standard Model, the Unsung Triumph of Modern Physics* (New York: Penguin, 2006).

6   L. Smolin, *Three Roads to Quantum Gravity: A New Understanding of Space, Time, and the Universe* (New York: Basic Books, 2001).

7   Quoted in F. S. Perls, *Gestalt Therapy Verbatim* (Gouldsboro, ME: Gestalt Journal Press, 1992).

8   Quoted in R. P. Feynman, *The Character of Physical Law* (New York: Penguin, 1992).

9   When Einstein first calculated the effect in 1911, he mistakenly calculated a deflection angle the same as for Newton's theory. Luckily for him and his reputation, an expedition planned for 1914 to watch starlight bend past the Sun during a solar eclipse was disrupted by the outbreak of World War I and observers already in place to watch the eclipse were captured by Russian soldiers. The correct deflection angle is twice the Newtonian value.

10 F. W. Dyson, A. S. Eddington, and C. Davidson, "A Determination of the Deflection of Light by the Sun's Gravitational Field, from Observations Made at the Total Eclipse of 29 May, 1919," *Philosophical Transactions of the Royal Society* 220A (1920): 291–333.

11 A. Calaprice, ed., *The New Quotable Einstein* (Princeton: Princeton University Press, 2005).

12   A. Einstein, "Lens-Like Action of a Star by the Deviation of Light in the Gravitational Field," *Science* 84 (1936): 506–07.

13   L. M. Krauss, "What Einstein Got Wrong," *Scientific American*, September 2015, 51–55.

14   F. Zwicky, "Nebulae as Gravitational Lenses," *Physical Review* 51 (1937): 290.

15   D. Walsh, R. F. Carswell, and R. J. Weymann, "0957+561 A, B: Twin Quasi-stellar Objects or Gravitational Lens?" *Nature* 279 (1979): 381–84.

16   The distance scale or the expansion rate of the universe is set by the slope of the relation between recession velocity and distance, $v = H_0 d$, where v is recession velocity, d is distance, and the slope of the relationship is the Hubble constant, H_0. Normally, the Hubble constant is measured by an overlapping chain of distance indicators, starting with parallax geometry for nearby stars, and extending through supernovas with well-calibrated peak brightness. Using gravitational lenses to measure the Hubble constant is direct and bypasses this entire chain of reasoning. Measuring a time delay in a lens system means the difference in distance between the two paths is measured. Since all the angles in the lens configuration are measured too, the entire geometry is determined, so giving the factor that relates distance and velocity or redshift.

17   J. N. Hewitt et al., "Unusual Radio Source MG 1131+0456: A Possible Einstein Ring?" *Nature* 333 (1988): 537–40.

18   There's a third form of gravitational lensing, in which light from distant galaxies is slightly distorted by all the dark matter along the line of sight. Think of the universe like a funhouse mirror where light doesn't travel in straight lines but undulates subtly due to the widely distributed dark matter. For an individual galaxy, the distortion is only 0.1%, too small to be detected, so it shows up when looking for patterns in the shapes of thousands of faint galaxies. For this reason it's referred to as statistical lensing. Statistical lensing demonstrates that space between galaxies is filled with dark matter.

19   U. I. Uggerhoj, R. E. Mikkelsen, and J. Faye, "The Young Center of the Earth," *European Journal of Physics* 37 (2016): 35602–10.

20   C. M. Will, "The Confrontation Between General Relativity and Experiment," *Living Reviews in Relativity* 9 (2006): 3–90.

21   R. V. Pound and G. A. Rebka, Jr., "Apparent Weight of Photons," *Physical Review Letters* 4 (1960): 337–41.

22   J. C. Hafele and R. E. Keating, "Around the World Atomic Clocks: Observed Relativistic Time Gains," *Science* 177 (1972): 168–70.

23   R. F. C. Vessot et al., "Test of Relativistic Gravitation with a Space-Borne Hydrogen Maser," *Physical Review Letters* 45 (1980): 2081–84.

24   H. Muller, A. Peters, and S. Chu, "A Precision Measurement of the Gravitational Redshift by Interference of Matter Waves," *Nature* 463 (2010): 926–29.

25   R. Wojtak, S. H. Hansen, and J. Hjorth, "Gravitational Redshift of Galaxies in Clusters as Predicted by General Relativity," *Nature* 477 (2011): 567–69.

26   L. Huxley, *The Life and Letters of Thomas Henry Huxley* (London: Mac-Millan, 1900), 189.

27   I. I. Shapiro et al., "Fourth Test of General Relativity: New Radar Result," *Physical Review Letters* 26 (1971): 1132–35.

28   B. Bertotti, L. Iess, and P. Tortora, "A Test of General Relativity using Radio Links with the Cassini Spacecraft," *Nature* 425 (2003): 374–76.

29   E. Teo, "Spherical Photon Orbits around a Kerr Black Hole," *General Relativity and Gravitation* 35 (2003): 1909–26.

30   For a rapidly spinning black hole, the innermost stable circular orbit might be inside the photon sphere, which means that material there is unobservable.

31   C. S. Reynolds and M. A. Nowak, "Fluorescent Iron Lines as a Probe of Astrophysical Black Hole Systems," *Physics Reports* 377 (2003): 389–466.

32   Y. Tanaka et al., "Gravitationally Redshifted Emission Implying an Accretion Disk and Massive Black Hole in the Active Galaxy MCG-6-30-15," *Nature* 375 (1995): 659–61.

33   J. F. Dolan, "Dying Pulse Trains in Cygnus XR-1: Evidence for an Event Horizon," *Publications of the Astronomical Society of the Pacific* 113 (2001): 974–82.

34   N. Shaposhnikov and L. Titarchuk, "Determination of Black Hole Masses in Galactic Black Hole Binaries Using Scaling of Spectral and Variability Characteristics," *Astrophysical Journal* 699 (2009): 453–68.

35   "Gravitational Vortex Provides New Way to Study Matter Close to a Black Hole," press release, European Space Agency, July 12, 2016, http://sci.esa .int/xmm-newton/58072-gravitational-vortex-provides-new-way-to-study-matter-close-to-a-black-hole/.

36   A. Ingram et al., "A Quasi-Periodic Modulation of the Iron Line Centroid Energy in the Black Hole Binary H1743-322," *Monthly Notices of the Royal Astronomical Society* 461 (2016): 1967–80.

37   M. Middleton, C. Done, and M. Gierlinski, "The X-Ray Binary Analogy to the First AGN QPO," *Proceedings of the AIP Conference on X-Ray Astronomy: Present Status, Multi-Wavelength Approaches, and Future Perspectives* 1248 (2010): 325–28.

38   M. J. Rees, "Tidal Disruption of Stars by Black Holes of $10^6 - 10^8$ Solar Masses in Nearby Galaxies," *Nature* 333 (1988): 523–28. This was the detailed development of an original idea from a decade earlier; see J. G. Hills, "Possible Power Source of Seyfert Galaxies and QSOs," *Nature* 254 (1975): 295–98.

39   S. Gezari, "The Tidal Disruption of Stars by Supermassive Black Holes," *Physics Today* 67 (2014): 37–42.

40   E. Kara, J. M. Miller, C. Reynolds, and L. Dai, "Relativistic Reverberation in the Accretion Flow of a Tidal Disruption Event," *Nature* 535 (2016): 388–90.

41   G. C. Bower, "The Screams of the Star Being Ripped Apart," *Nature* 351 (2016): 30–31.

42   G. Ponti et al., "Fifteen Years of XMM-Newton and Chandra Monitoring of Sgr A°: Evidence for a Recent Increase in the Bright Flaring Rate," *Monthly Notices of the Royal Astronomical Society* 454 (2015): 1525–44.

43   Jacob Aron, "Black holes devour stars in gulps and nibbles," *New Scientist*,

March 25, 2015, https://www.newscientist.com/article/mg22530144-400-black-holes-devour-stars-in-gulps-and-nibbles/.

44   Richard Gray, "Echoes of a stellar massacre," *Daily Mail*, September 16, 2016, http://www.dailymail.co.uk/sciencetech/article-3793042/Echoes-stellar-massacre-Gasps-dying-stars-torn-apart-supermassive-black-holes-detected.html.

45   C. W. F. Everitt, "The Stanford Relativity Gyroscope Experiment: History and Overview," in *Near Zero: Frontiers in Physics*, edited by J. D. Fairbank et al. (New York: W.H. Freeman, 1989).

46   Gravity Probe B is a great example of the perseverance and technology development required for many space missions. The concept stemmed from a theoretical paper written by Stanford professor Leonard Schiff in 1957. He and MIT professor George Pugh proposed the mission to NASA in 1961 and the project received its first funding in 1964. There followed forty years of technology development and delays caused by NASA's Shuttle program. Schiff and Pugh died long before the launch in 2004.

47   C. W. F. Everitt et al., "Gravity Probe B: Final Results of a Space Experiment to Test General Relativity," *Physical Review Letters* 106 (2011): 22101–06.

48   E. S. Reich, "Spin Rate of Black Holes Pinned Down," *Nature* 500 (2013): 135.

49   K. Middleton, "Black Hole Spin: Theory and Observations," in *Astrophysics of Black Hole, Astrophysics and Space Science Library,* volume 440 (Berlin, Springer, 2016), 99–137.

50   J. W. T. Hessels et al., "A Radio Pulsar Spinning at 716 Hz," *Science* 311 (2006): 1901–04.

51   L. Gou et al., "The Extreme Spin of the Black Hole in Cygnus X-1," *Astrophysical Journal* 742 (2011): 85–103.

52   M. J. Valtonen, "Primary Black Hole Spin in OJ 287 as Determined by the General Relativity Centenary Flare," *Astrophysical Journal Letters* 819 (2016): L37–43.

53   Quoted in Dennis Overbye, "Black Hole Hunters," *New York Times*, June 8, 2015, http://www.nytimes.com/2015/06/09/science/black-hole-event-horizon-telescope.html.

54   A. Ricarte and J. Dexter, "The Event Horizon Telescope: Exploring Strong Gravity and Accretion Physics," *Monthly Notices of the Royal Astronomical Society* 446 (2014): 1973–87.

55   S. Doeleman et al., "Event-Horizon-Scale Structure in the Supermassive Black Hole Candidate at the Galactic Center," *Nature* 455 (2008): 78–80.

56   T. Johannsen et al., "Testing General Relativity with the Shadow Size of SGR A°," *Physical Review Letters* 116 (2016): 031101.

7: SEEING WITH GRAVITY EYES

1   F. G. Watson, *Stargazer: The Life and Times of the Telescope* (Cambridge, MA: De Capo Press, 2005).

2   P. Morrison, "On Gamma-Ray Astronomy," *Il Nuovo Cimento* 7 (1958): 858–65.

3   Four prominent examples are A. A. Abdo et al., "Fermi-LAT Observations of Markarian 421: the Missing Piece of its Spectral Energy Distribution," *Astrophysical Journal* 736 (2011): 131–53; V. A. Acciari et al., "The Spectral Energy Distribution of Markarian 501: Quiescent State Versus Extreme Outburst," *Astrophysical Journal* 729 (2011): 2–11; V. S. Paliya,"A Hard Gamma-Ray Flare from 3C 279 in December 2013," *Astrophysical Journal* 817 (2016): 61–75; and S. Soldi et al., "The Multiwavelength Variability of 3C 273," *Astronomy and Astrophysics* 486 (2008): 411–27.

4   For the sake of the analogy, let's momentarily suspend our disbelief, take a materialist view of mind and brain, and imagine that one day we can use remote sensing to parse thoughts.

5   Gravitational waves do not occur, however, when the motion is perfectly symmetric, like an expanding or contracting sphere, or rotationally symmetric, like a spinning disk or sphere. A perfectly symmetric supernova collapse or a perfectly spherical spinning neutron star would not emit gravitational waves. To put it technically, the third time derivative of the quadrupole moment in the stress-energy tensor must be nonzero for a system to emit gravitational radiation. In mathematical terms, this is analogous to the changing dipole moment of charge or current that leads to electromagnetic radiation. Got it?

6   P. G. Bergmann, *The Riddle of Gravitation* (New York: Charles Scribner's Sons, 1968).

7   It's an assumption and a supposition that gravity and gravitational waves propagate at the speed of light. No experiment to test this has ever been unequivocally successful. It's very difficult to design any experiment to "turn off" gravity or change it dramatically enough at a remote location to see how fast it travels. In the standard model of particle physics, gravity is carried by a particle called the graviton, traveling at light speed. Gravitons have never been detected.

8   A. S. Eddington, "The Propagation of Gravitational Waves," *Proceedings of the Royal Society of London* 102 (1922): 268–82.

9   K. Daniel, "Einstein versus the *Physical Review*," *Physics Today* 58 (2005): 43–48.

10   A. Einstein and N. Rosen, "On Gravitational Waves," *Journal of the Franklin Institute* 223 (1937): 43–54

11   Gravity Research Foundation website, http://www.gravityresearchfoundation .org/origins.html.

12   I won't belabor this with economics references in an astronomy book, but there is an extensive literature to show that, while timing markets can work in certain sectors and for short periods of time, as a long-term strategy it is ruinous. Babson was simply lucky; it happens.

13   J. L. Cervantes-Cota, S. Galindo-Uribarri, and G. F. Smoot, "A Brief History of Gravitational Waves," *Universe* 2 (2016): 22–51.

14   M. Gardner, *Fads and Fallacies in the Name of Science* (New York: Dover, 1957), 93.

15   Despite its origin in pseudoscience and magical thinking, Babson's vision
      was in the end very positive. In time, the Gravity Research Foundation
      recaptured prestige in the eyes of the physics community. The 1957 confer-
      ence in Chapel Hill is today known as the GR1 Conference. It began as a
      series of international conferences every few years to discuss the state of the
      art in gravitation and general relativity. As an indication of the international
      nature of the field, the last seven meetings were held in India, South Africa,
      Ireland, Australia, Mexico, Poland, and most recently, GR21 in New York.

16   Janna Levin, "Gravitational Wave Blues," https://aeon.co/essays/how-joe-
      weber-s-gravity-ripples-turned-out-to-be-all-noise.

17   Weber's concept was published in J. Weber, "Detection and Generation
      of Gravitational Waves," *Physical Review* 117 (1960): 306–13. The per-
      formance of his first operating detector was published six years later in J.
      Weber, "Observations of the Thermal Fluctuations of a Gravitational-Wave
      Detector," *Physical Review Letters* 17 (1966): 1228–30.

18   J. Weber, "Evidence for Discovery of Gravitational Radiation" *Physical
      Review Letters* 22 (1969): 1320–24, followed closely by J. Weber, "Anisot-
      ropy and Polarization in the Gravitational-Radiation Experiments," *Physical
      Review Letters* 25 (1970): 180–84.

19   I never met Weber, but I know his wife, Virginia Trimble, well. She's a fellow
      Brit and an expert on the history of astronomy, so we exchange astronomy
      arcana occasionally. In their long marriage, Virginia had a faculty job at the
      University of California in Irvine, so she spent half her year there and half
      back East where Weber had his faculty job. After he died in 2000, we met
      at a conference and talked about his work, and I could tell it was a painful
      subject. She had to watch him be denigrated and belittled by people who
      had no idea how hard he had worked to hone his technique. He continued
      his research for over twenty years after federal support was withdrawn. Vir-
      ginia said it took a severe toll on him, emotionally and physically.

20   J. A. Wheeler, *Geons, Black Holes, and Quantum Foam: A Life in Physics*
      (New York: Norton, 1998), 257–58.

21   J. M. Weisberg, D. J. Nice, and J. H. Taylor, "Timing Measurements of the
      Relativistic Binary Pulsar PSR B1913+16," *Astrophysical Journal* 722 (2010):
      1030–34.

22   The binary system emits 7×10^{24} watts of gravitational radiation, and the dis-
      tance between the two neutron stars shrinks by 3.5 meters per year. It will
      take 300 million years for the two neutron stars to collide and merge. Even the
      Solar System emits gravitational radiation, but far, far less, only 5,000 watts.

23   This is speculation, informed by the properties of the gravity waves detected
      when the black holes merged and by plausible formation scenarios that could
      lead to black holes this massive—more massive than any black holes in the
      local universe. Massive stars forming 11 billion years ago would have a much
      smaller proportion of heavy elements than the Sun, and models suggest their
      initial mass could be higher than stars forming now. As a result, these ancient
      stars would shed less mass and leave behind more massive black holes. This sce-

nario is described in K. Belczynski, D.E. Holz, T. Bulik, and R. O'Shaughnessy, "The First Gravitational-Wave Source from the Isolated Evolution of Two Stars in the 40–100 Solar Mass Range," *Nature* 534 (2016): 512–15. A more radical possibility, not ruled out by the data, is that the black holes were primordial, formed in the early universe from dark matter; see S. Bird et al., "Did LIGO Detect Dark Matter," *Physical Review Letters* 116 (2016): 201301–07.

24   J. Chu, "Rainer Weiss on LIGO's Origins," oral history, Massachusetts Institute of Technology Q & A News series, http://news.mit.edu/2016/rainer-weiss-ligo-origins-0211.

25   Weiss credits his students and also Phillip Chapman, an MIT researcher who went to work for NASA and then stopped working on gravity and physics. Intriguingly, and ironically, the antecedent for the interferometer was Joseph Weber, who suggested the idea to his former student Robert Forward in 1964. Forward used funds from his employer, Hughes Research Lab, to build a proto-type interferometer with 8.5-meter-long arms. After 150 hours of observations, he detected nothing. Confirming the "small world" nature of the gravitational physics community, Forward credited conversations with Rainer Weiss in a footnote to his paper, R. L. Forward, "Wide-Band Laser-Interferometer Gravitational-Radiation Experiment," *Physical Review D* 17 (1978): 379–90.

26   R. Weiss, "Quarterly Progress Report, Number 102, 54-76," Research Laboratory of Electronics, MIT, 1972, http://dspace.mit.edu/bitstream/handle/1721.1/RLE_QPR_105_V.pdf?sequence=1.

27   Quoted in J. Levin, *Black Hole Blues and Other Songs from Outer Space* (New York: Knopf, 2016).

28   Quoted in N. Twilley, "Gravitational Waves Exist: The Inside Story of How Scientists Finally Found Them," *New Yorker*, February 11, 2016, http://www.newyorker.com/tech/elements/gravitational-waves-exist-heres-how-scientists-finally-found-them.

29   More specifically, they were the best in the United States. In this LIGO-centric narrative, I omit for simplicity the substantial early efforts of other groups and other countries. Drever's group at the University of Glasgow kept up their work on interferometers after he left for Caltech. Meanwhile, a group led by Peter Kafka in Germany learned about Weiss's work in 1974 and hired one of his students to build interferometers. They collaborated with an Italian group to build 3-meter and 30-meter prototypes over the next decade. Interestingly, in a demonstration of the "small world" phenomena of gravitational wave research, Drever had first learned about interferometers at a lecture by Peter Kafka in 1975. The German group and the Scottish group combined to propose a kilometer-scale instrument in the mid-1980s but it wasn't funded. Eventually they were able to build a 600-meter instrument, which started operations in 2001 and was a critical test-bed for LIGO detectors and techniques. The French had ideas for an even more ambitious interferometer, led by Alain Brillet, who had worked with Weiss at MIT in the early 1980s. The Virgo project started taking data in 2004 and has been in a full partnership with LIGO for a decade. The details of worldwide

efforts to detect gravitational waves are given in J. L. Cervantes-Cota, S. Galindo-Uribarri, and G. F. Smoot, "A Brief History of Gravitational Waves," *Universe* 2 (2016): 22–51.

30   P. Linsay, P. Saulson, and R. Weiss, "A Study of a Long Baseline Gravitational Wave Antenna System," 1983, https://dcc.ligo.org/public/0028/T830001/000/NSF_bluebook_1983.pdf.

31   LIGO reports and newsletters don't convey these tensions. They understandably have a mostly valedictory tone given the ultimate success of the project. The best insider-outsider account is contained in Janna Levin's book *Black Hole Blues and Other Songs from Outer Space* (New York: Knopf, 2016).

32   A. Cho, "Here is the First Person to Spot Those Gravitational Waves," *Science*, February 11, 2016, http://www.sciencemag.org/news/2016/02/here-s-first-person-spot-those-gravitational-waves.

33   Quoted in Josh Rottenberg, "Meet the Astrophysicist Whose 1980 Blind Date Led to *Interstellar*," *Los Angeles Times*, November 21, 2014, http://www.latimes.com/local/great-reads/la-et-c1-kip-thorne-interstellar-20141122-story.html.

34   Academic lineages exist in all fields but they are particularly strong in theoretical physics and mathematics. A career can be shaped and launched by having the right thesis advisor and students who reflect well on their advisors. In theoretical fields, the influence of an advisor can extend to "taste" in choosing a problem to solve and the "style" with which it is solved. These aesthetic considerations are often opaque to an outsider. Kip Thorne has mentored fifty PhD students as a professor at Caltech, including many influential figures in theoretical astrophysics and relativity such as Alan Lightman, Bill Press, Don Page, Saul Teukolsky, and Clifford Will.

35   "How Are Gravitational Waves Detected?" Q & A with Rainer Weiss and Kip Thorne, *Sky and Telescope*, August 28, 2016, http://www.skyandtelescope.com/astronomy-resources/astronomy-questions-answers/science-faq-answers/kavli-how-gravitational-waves-detected/.

36   K. S. Thorne, *Black Holes and Time Warps: Einstein's Outrageous Legacy* (New York: W. W. Norton, 1994).

37   See Adam Rogers, "Wrinkles in Spacetime: The Warped Astrophysics of Interstellar," *Wired*, https://www.wired.com/2014/10/astrophysics-interstellar-black-hole/

38   J. Updike, "Cosmic Gall," *New Yorker*, December 17, 1960, 36.

39   K. S. Thorne, "Gravitational Radiation," in *Three Hundred Years of Gravitation*, edited by S. Hawking and W. W. Israel (Cambridge: Cambridge University Press, 1987), 330–458.

40   This information is laid out clearly and graphically in the *LIGO Magazine*, no. 8, March 2016, http://www.ligo.org/magazine/LIGO-magazine-issue-8.pdf.

41   This will be a critical advance, as it's impossible to identify the sources of LIGO black hole signals so far. Gravitational waves represent a new way of looking at the universe so it's frustrating not to be able to identify the objects responsible and observe them with light and across the electromagnetic

spectrum. There are other details of the detection process that affect the interpretation of the data. Interferometers are most sensitive to the waves that arrive from above, since they are stretching and squeezing within the transverse plane. At any other angle the signal is less. With two detectors separated by thousands of miles, they are not co-planar due to the curvature of the Earth, so that must be taken into account too. The signal is largest for a binary orbit with a plane that faces the Earth, and lower for other inclinations. LIGO experimenters must extract every iota of information possible from each of their transient events.

42   In terms of the peculiar arithmetic that applies to merging black holes, the first event involved the sum 36 + 29 = 62 solar masses, with 3 solar masses emitted as gravitational waves. The second event involved the sum 14 + 9 = 21 solar masses, with 2 solar masses emitted as gravitational waves, and the "candidate" event involved the sum 23 + 13 = 34 solar masses, with 2 solar masses emitted as gravitational waves. The detection significance of the three events was >5.3σ for the first two, and a marginal 1.7σ for the candidate event. Localization on the sky depends on signal strength; it was 230 square degrees for the first event, 850 square degrees for the second event, and 1600 square degrees for the candidate event. In general, the characteristic chirp frequency scales with the black hole mass as $M^{-5/8}$ and the displacement in the interferometer, h, scales with the black hole mass as $M^{5/3}$. All these measurements, and more, are in the *LIGO Magazine*, no. 9, August 2016, http://www.ligo.org/magazine/LIGO-magazine-issue-9.pdf.

43   A. Murguia-Merthier et al., "A Neutron Star Binary Merger Model for GW170817/GRB 170817A/SSS17a," *Astrophysical Journal Letters* 848 (2017): L34–42.

44   M. R. Seibert et al., "The Unprecedented Properties of the First Electromagnetic Counterpart to a Gravitational-Wave Source," *Astrophysical Journal Letters* 848 (2017): L26–32.

45   J. Abadie et al., "Predictions for the Rates of Compact Binary Coalescences Observable by Ground-Based Gravitational-Wave Detectors," *Classical Quantum Gravity* 27 (2010): 173001–26.

46   B. P. Abbott et al., "The Rate of Binary Black Hole Mergers Inferred from Advanced LIGO Observations Surrounding GW150914," *Astrophysical Journal Letters* 833 (2016): L1–99. Advanced LIGO working in conjunction with the VIRGO interferometer in Europe will deliver source positions of 5 square degrees, 100 times more accurate than the early LIGO detections.

47   LISA was originally a joint project of NASA and ESA. Initial design studies date back to the 1980s. But NASA ran into budget problems and withdrew from the project in 2011, so ESA went from being a partner to being the sole agency sponsoring this ambitious mission. LISA is a major new mission in ESA's "Cosmic Vision" program, with a tentative launch date of 2034. See https://www.elisascience.org/news/top-news/gravitationaluniverseselecteda sl3.

48   M. Armano et al., "Sub-Femto-g Free Fall for Space-Based Gravitational Wave Observatories: LISA Pathfinder Results," *Physical Review Letters* 116 (2016): 231101–11.

49   Analogous to the situation for stellar mass black holes, the most difficult problem to understand is the timescale of the final merger. The difficulty of the supermassive black holes losing enough angular momentum to merge is called the "final parsec" problem. In a gas-rich galaxy the final merger phase might take 10 million years, but in a gas-poor galaxy it might take billions of years. In some models, it might take longer than the age of the universe, meaning that massive galaxies might contain binary supermassive black holes that have never merged—which, in turn, would mean that there would be no gravitational wave signal to detect.

50   J. Salcido et al., "Music from the Heavens: Gravitational Waves from Super-massive Black Hole Mergers in the EAGLE Simulations," *Monthly Notices of the Royal Astronomical Society* 463 (2016): 870–85.

51   G. Hobbs, "Pulsars as Gravitational Wave Detectors," in *High Energy Emission from Pulsars and Their Systems*, Astrophysics and Space Science Proceedings (Berlin: Springer, 2011), 229–40.

52   S. R. Taylor et al., "Are We There Yet? Time to Detection of Nano-Hertz Gravitational Waves Based on Pulsar-Timing Array Limits," *Astrophysical Journal Letters* 819 (2016): L6–12.

53   A. Guth, *The Inflationary Universe: The Quest for a New Theory of Cosmic Origins* (New York: Perseus, 1997).

54   P. D. Lasky et al., "Gravitational Wave Cosmology Across 29 Decades in Frequency," *Physical Review X* 6 (2016): 011035–46.

55   Technically, this pattern is called B-mode polarization. It means that the electromagnetic field has a pattern like a vortex superimposed. The temperature of the microwaves is uniform across the sky to one part in 100,000, and the polarization signal is 100 times smaller, so detecting the gravitational wave effect requires an extraordinary level of precision.

56   D. Hanson et al., "Detection of B-Mode Polarization in the Cosmic Microwave Background with Data from the South Pole Telescope," *Physical Review Letters* 111 (2014): 141301–07.

8: THE FATE OF BLACK HOLES

1   Fermions are half-integer spin particles that obey statistics defined by Enrico Fermi and Paul Dirac in the 1930s. No two fermions can have exactly the same set of quantum properties. Fundamental fermions include the electron and the six types of quark. Composite fermions include protons and neutrons. Bosons are integer spin particles that obey statistics defined by Albert Einstein and Satyendra Bose in the 1920s. Fundamental bosons include the photon, the Higgs boson, and the (still hypothetical) graviton. Composite bosons include the helium nucleus and the carbon nucleus. Any number of

bosons can have the same quantum state. While fermions are thought of as particles and bosons as force carriers, the distinction between those two categories in quantum mechanics is not clear-cut.

2 Note that the idea of extra dimensions is not necessarily a reason to doubt string theory as a description of nature. The mathematics of multidimensional spaces were worked out in the middle of the nineteenth century by Gauss and Bolyai. In the 1920s, Kaluza and Klein did early work on a theory of gravity that incorporated an extra dimension. String theory is still a very active field of theoretical physics, and progress has been made, but there's also been a backlash. For the positive view of string theory's beauty and potential as a theory of everything, see B. Greene, *The Elegant Universe: Superstrings, Hidden Dimensions, and the Quest for the Ultimate Theory* (New York: W. W. Norton, 2003). For a countervailing view, see L. Smolin, *The Trouble with Physics: The Rise of String Theory, the Fall of a Science, and What Comes Next* (New York: Houghton Mifflin, 2006).

3 In a nonrotating black hole the singularity is a point, and in a rotating black hole it is a ring. To a physicist a ringlike singularity is no less distasteful than a pointlike singularity because it still has infinite space-time curvature at every point along its circumference.

4 J. Womersley, "Beyond the Standard Model," *Symmetry*, February 2005, 22–25. A slightly more technical article with the same title is J. D. Lykken, "Beyond the Standard Model," a lecture given at the 2009 European School of High Energy Physics, *CERN Yellow Report CERN-2010-0002* (Geneva: CERN, 2011), 101–09.

5 L. Randall and R. Sundrum, "An Alternative to Compactification," *Physical Review Letters* 83 (1999): 4690–93.

6 L. Randall, *Warped Passages: Unraveling the Mysteries of the Universe's Hidden Dimensions* (New York: Ecco, 2005).

7 M. Holloway, "The Beauty of Branes," *Scientific American* 293, November 2005, 38-40.

8 L. Randall, "Theories of the Brane," in *The Universe: Leading Scientists Explore the Origin, Mysteries, and Future of the Cosmos*, edited by J. Brockman (New York: HarperCollins, 2014), 62–78.

9 e. e. cummings, "Pity this busy monster, manunkind," in *e. e. cummings: Complete Poems 1904–1962* (New York: W. W. Norton, 1944).

10 J. Neilsen et al., "The 3 Million Second Chandra Campaign on Sgr A°: A Census of X-ray Flaring Activity from the Galactic Center," in *The Galactic Center: Feeding and Feedback in a Normal Galactic Nucleus*, Proceedings of the International Astronomical Union, vol. 303 (2013): 374–78.

11 M. Nobukawa et al., "New Evidence for High Activity of the Super-Massive Black Hole in our Galaxy," *Astrophysical Journal Letters* 739 (2011): L52–56.

12 F. Nicastro et al., "A Distant Echo of Milky Way Central Activity Closes the Galaxy's Baryon Census," *Astrophysical Journal Letters* 828 (2016): L12–20.

13 'Chandra Finds Evidence for Swarm of Black Holes Near the Galactic Center," NASA press release, January 10, 2005, http://chandra.harvard.edu/press/05_releases/press_011005.html.

14   D. Haggard et al., "The Field X-ray AGN Fraction to z = 0.7 from the Chandra Multi-Wavelength Project and the Sloan Digital Sky Survey," *Astrophysical Journal* 723 (2010): 1447–68.

15   R. P. van der Marel et al., "The M31 Velocity Vector: III. Future Milky Way-M31-M33 Orbital Evolution, Merging, and Fate of the Sun," *Astrophysical Journal* 753 (2012): 1–21.

16   T. J. Cox and A. Loeb, "The Collision Between the Milky Way and Andromeda," *Monthly Notices of the Royal Astronomical Society* 386 (2007): 461–74.

17   Study of M31 is complicated by the fact that it has a double nucleus within a dense star cluster. The brighter of the two concentrations is offset from the center of the galaxy, and the fainter one, 5 light years away, contains the massive black hole. The distance of 2.5 million light years makes the nuclear regions difficult to study in detail even with the Hubble Space Telescope. The best measurement of the black hole mass is in the range 110 to 230 million solar masses. See R. Bender et al., "HST STIS Spectroscopy of the Triple Nucleus of M31: Two Nested Disks in Keplerian Rotation Around a Supermassive Black Hole," *Astrophysical Journal* 631 (2005): 280–300.

18   J. Dubinski, "The Great Milky Way-Andromeda Collision," *Sky and Telescope*, October 2006, 30–36. A more technical treatment is F. M. Khan et al., "Swift Coalescence of Supermassive Black Holes in Cosmological Mergers of Massive Galaxies," *Astrophysical Journal* 828 (2016): 73–80. The theory of how the final merger takes place is uncertain; see M. Milosavljevic and D. Merritt, "The Final Parsec Problem," in *The Astrophysics of Gravitational Wave Sources*, AIP Conference Proceedings, vol. 686 (2003): 201–10.

19   F. Khan et al, "Swift Coalescence of Supermassive Black Holes in Cosmological Mergers of Massive Galaxies," *Astrophysical Journal* 828 (2016): 73–81.

20   T. Liu et al., "A Periodically Varying Luminous Quasar at z = 2 from the PAN-STARRS1 Medium Deep Survey: A Candidate Supermassive Black Hole in the Gravitational Wave-Driven Regime," *Astrophysical Journal Letters* 803 (2015): L16–21.

21   K. Thorne, *The Science of Interstellar* (New York: W. W. Norton, 2014).

22   W. Zuo et al., "Black Hole Mass Estimates and Rapid Growth of Supermassive Black Holes in Luminous z = 3.5 Quasars," *Astrophysical Journal* 799 (2014): 189–201.

23   G. Ghisellini et al., "Chasing the Heaviest Black Holes of Jetted Active Galactic Nuclei," *Monthly Notices of the Royal Astronomical Society* 405 (2010): 387–400.

24   K. Inayoshi and Z. Haiman, "Is There a Maximum Mass for Black Holes in Galactic Nuclei?," *Astrophysical Journal* 828 (2016): 110–17.

25   D. Sobral et al., "Large H-Alpha Survey at z = 2.23, 1.47, 0.84, and 0.40: The 11 Gyr Evolution of Star-forming Galaxies from HiZELS," *Monthly Notice of the Royal Astronomical Society* 428 (2013): 1128–46.

26   F. C. Adams and G. Laughlin, "A Dying Universe: The Long Term Fate and Evolution of Astrophysical Objects," *Reviews of Modern Physics* 69 (1997): 337–72.

27   A. Burgasser, "Brown Dwarfs: Failed Stars, Super Jupiters," *Physics Today*, June 2008, 70–71.

28　　D. N. Spergel, "The Dark Side of Cosmology: Dark Matter and Dark Energy," *Science* 347 (2015): 1100–02.

29　　Astronomers have wondered how future inhabitants of Milkomeda would know they lived in an expanding universe if there were no galaxies visible by which to measure redshifts. After a trillion years the expansion will have progressed so far that the microwaves left over from the big bang will have left the event horizon. It seems that the only evidence of the universe beyond Milkomeda will be hypervelocity stars continuously being ejected from Milkomeda and all other galaxies at close to the speed of light. This possibility is described in A. Loeb, "Cosmology with Hypervelocity Stars," *Journal of Cosmology and Astroparticle Physics* 4 (2011): 23–29.

30　　F. Adams and G. Laughlin, *The Five Ages of the Universe* (New York: Free Press, 1999).

31　　H. Nishino, Super-K Collaboration, "Search for Proton Decay in a Large Water Cerenkov Detector," *Physical Review Letters* 102 (2012): 141801–06.

32　　J. Baez, "The End of the Universe," http://math.ucr.edu/home/baez/end.html

33　　W. B. Yeats, "The Second Coming" (1919), in *The Classic Hundred Poems* (New York: Columbia University Press, 1998).

34　　A. Eddington, *The Nature of the Physical World: Gifford Lectures of 1927* (Newcastle-upon-Tyne: Cambridge Scholars, 2014).

35　　B. W. Jones, *Life in Our Solar System and Beyond* (Berlin: Springer, 2013).

36　　The Extrasolar Planets Encyclopedia is continuously updated, http://exoplanet.eu/.

37　　R. Jayawardhana, *Strange New Worlds: The Search for Alien Planets and Life Beyond our Solar System* (Princeton: Princeton University Press, 2013).

38　　A. Cassan et al., "One or More Bound Planets per Milky Way Star from Microlensing Observations," *Nature* 481 (2012): 167–69.

39　　F. J. Dyson, "Time Without End: Physics and Biology in an Open Universe," *Reviews of Modern Physics* 51 (1979): 447–60.

40　　M. Bhat, M. Dhurandhar, and N. Dadhich, "Energetics of the Kerr-Newman Black Hole by the Penrose Process," *Journal of Astronomy and Astrophysics* 6 (1985): 85–100.

41　　T. Opatrny, L. Richterek, and P. Bakala, "Life Under a Black Sun," 2016, https://arxiv.org/abs/1601.02897.

42　　F. J. Dyson, "Search for Artificial Stellar Sources of Infra-Red Radiation," *Science* 131 (1960): 1667–68.

INDEX

Note: Page numbers in *italics* indicate illustrations.